T0269476

Learning from the Impacts of Superstorm Sandy

Learning from the Impacts of Superstorm Sandy

J. Bret Bennington
E. Christa Farmer

AMSTERDAM • BOSTON • HEIDELBERG • LONDON
NEW YORK • OXFORD • PARIS • SAN DIEGO
SAN FRANCISCO • SINGAPORE • SYDNEY • TOKYO

Academic Press is an imprint of Elsevier

Academic Press is an imprint of Elsevier
32 Jamestown Road, London NW1 7BY, UK
The Boulevard, Langford Lane, Kidlington, Oxford OX5 1GB, UK
Radarweg 29, PO Box 211, 1000 AE Amsterdam, The Netherlands
225 Wyman Street, Waltham, MA 02451, USA
525 B Street, Suite 1900, San Diego, CA 92101-4495, USA

Notices
Knowledge and best practice in this field are constantly changing. As new research
and experience broaden our understanding, changes in research methods, professional
practices, or medical treatment may become necessary.

Practitioners and researchers must always rely on their own experience and knowledge in
evaluating and using any information, methods, compounds, or experiments described
herein. In using such information or methods they should be mindful of their own safety
and the safety of others, including parties for whom they have a professional responsibility.

To the fullest extent of the law, neither the Publisher nor the authors, contributors, or
editors, assume any liability for any injury and/or damage to persons or property as a
matter of products liability, negligence or otherwise, or from any use or operation of any
methods, products, instructions, or ideas contained in the material herein.

British Library Cataloguing in Publication Data
A catalogue record for this book is available from the British Library

Library of Congress Cataloging-in-Publication Data
A catalog record for this book is available from the Library of Congress

ISBN: 978-0-12-801520-9

For information on all Academic Press publications
visit our website at http://store.elsevier.com/

Working together
to grow libraries in
developing countries

www.elsevier.com • www.bookaid.org

CONTENTS

LIST OF CONTRIBUTORS

Riley Behrens

New York Water Science Center, United States Geological Survey, Coram, NY, USA

Alan I. Benimoff

Department of Engineering Science and Physics, College of Staten Island/CUNY, Staten Island, NY, USA

J. Bret Bennington

Department of Geology, Environment and Sustainability, Hofstra University, Hempstead, NY, USA

Andrew Coburn

Program for the Study of Developed Shorelines, Western Carolina University, Cullowhee, NC, USA

Elizabeth Denis

Department of Geosciences, Pennsylvania State University, University Park, PA, USA

E. Christa Farmer

Department of Geology, Environment and Sustainability, Hofstra University, Hempstead, NY, USA

Xiahong Feng

Department of Earth Sciences, Dartmouth College, Hanover, NH, USA

Ezgi Akpinar Ferrand

Department of Geography, Southern Connecticut State University, New Haven, CT, USA

Katherine Freeman

Department of Geosciences, Pennsylvania State University, University Park, PA, USA

William J. Fritz

CUNY Graduate Center, College of Staten Island, Staten Island, NY, USA

Bowen Gabriel

Department of Geology and Geophysics, University of Utah, Salt Lake City, UT 84112, USA

Stephen Good

Department of Geology and Geophysics, University of Utah, Salt Lake City, UT, USA

Adam D. Griffith

Department of Geography, UNC – Charlotte, Geography Department, Charlotte, NC, USA

Michael Kress

College of Staten Island, Staten Island, NY, USA

Shuning Li
Department of Earth
and Planetary Science,
Johns Hopkins University,
Baltimore, MD, USA

John Lin
Department of Atmospheric
Sciences, University of Utah, Salt
Lake City, UT, USA

Derek Mallia
Department of Atmospheric
Sciences, University of Utah, Salt
Lake City, UT, USA

Katie Peek
Program for the Study of Developed
Shorelines, Western Carolina
University, Cullowhee, NC, USA

Amy Simonson
New York Water Science Center,
United States Geological Survey,
Coram, NY, USA

James Tait
Department of Science Education
and Environmental Studies, Southern
Connecticut State University,
New Haven, CT, USA

Robert Young
Program for the Study of Developed
Shorelines, Western Carolina
University, Cullowhee, NC, USA

Nicolas Zegre
Foresty and Natural Resources,
West Virginia University,
Morgantown, WV, USA

CHAPTER *1*

Introduction

J. Bret Bennington and E. Christa Farmer

ABSTRACT

"Superstorm" (hurricane turned extratropical cyclone) Sandy was one of the largest storms ever recorded in the Atlantic. Sandy brought record-breaking storm surge and coastal flooding to the area from southern New Jersey to eastern Long Island, resulting in the second costliest storm in U.S. history after Hurricane Katrina. Sandy's large size and severe impacts on coastal areas brought tragedy and suffering to the New York metropolitan region, and served as a wake-up call to better understand and prepare for the impacts of future storms exacerbated by the effects of climate change. The seven research articles in this volume are the first wave of the scientific response to this call.

Keywords: Superstorm Sandy; hurricane Sandy; extratropical cyclone Sandy

Superstorm Sandy was born as a tropical wave over the eastern Atlantic that developed into a tropical storm in the Caribbean Sea on October 22, 2012, eventually reaching Category 3 strength as Hurricane Sandy made landfall in Cuba on October 25. Passing through the Bahamas Sandy weakened, briefly becoming a tropical storm again, before regaining Category 1 strength as it moved northward parallel to the southeastern United States coast. As Sandy became extratropical approaching landfall just northeast of Atlantic City, New Jersey, there was a sense of complacency among the public. Only a little more than one year earlier, Hurricane Irene had come ashore on the New Jersey coast and again in Brooklyn as a tropical storm, causing catastrophic flooding in eastern New York State and central and southern Vermont, but inflicting relatively moderate damage in New York City and coastal Long Island (Lixion and Cangialosi, 2011). People who had evacuated for Irene, only to experience little to no flooding, stayed home for Sandy. However, the

Learning from the Impacts of Superstorm Sandy. http://dx.doi.org/10.1016/B978-0-12-801520-9.00001-8

official warnings for Sandy were dire, as the hurricane experts cautioned that this was no ordinary storm.

Hurricane turned extratropical cyclone ("Superstorm") Sandy was one of the largest storms ever recorded in the Atlantic, with a gale force wind field that extended for as far as 870 nautical miles at the peak of storm development (Blake et al., 2013). This extraordinarily large wind field, combining with a westward turn of the storm track toward the central New Jersey coast, pushed high waves and water into the funnel-shaped New York bight, resulting in record-breaking storm surge and coastal flooding from southern New Jersey to eastern Long Island, with the most severe impacts along the northern New Jersey coast, Staten Island, Lower Manhattan, and southwestern Long Island. Coastal communities in New Jersey, Staten Island, and Long Island were devastated and lower Manhattan was brought to a standstill by submerged roads, tunnels, and subway stations. Over 650,000 houses were damaged or destroyed in the U.S., most by waves and storm surge, and 72 fatalities resulted in the U.S. from the direct effects of the storm, including deaths from drowning (41) and falling tree limbs (20) (Blake et al., 2013; Rappaport, 2014). Overall, storm related costs exceeded 50 billion dollars making Sandy the second costliest storm in U.S. history, second only to hurricane Katrina. With Sandy, the "worst case" scenario storm for the New York region finally arrived, the one that scientists and emergency preparedness professionals had been warning about for decades.

Both of us (Farmer and Bennington) as well as several other contributors to this volume experienced Superstorm Sandy directly, either as residents of the New York metropolitan region or as researchers sent here to deploy instrumentation to study the storm. We were direct witnesses to the impacts and aftermath of Sandy and in many cases uniquely positioned to take advantage of the research opportunities provided by this unfortunate but scientifically opportune event. As one of our colleagues in the biology department at Hofstra University remarked while surveying the devastated trees in his neighborhood, "Wow, look at all this data!"

The seven research articles in this volume represent the first wave of studies to come from the data generated by Superstorm Sandy. Chapters 2 ("Measuring Storm Tide and High-water Marks from Hurricane Sandy in New York" by Amy Simonson and Riley Behrens) and 3 ("Superstorm Sandy and Staten Island: Learning from the Past,

Preparing for the Future" by Alan Benimoff and others) compile and analyze data and models of storm surge from Superstorm Sandy. Simonson and Behrens find that surge heights were greatest where the surge was augmented by coinciding with astronomical high tide reaching between 12 and nearly 20 ft in Richmond and Kings Counties. This agrees with modeling studies performed by Benimoff and others using actual Sandy data, which calculated surges up to 3.8 m (12.5 ft) in Richmond County. Refining these models will help better predict the impacts of future storms of varying size, strength, track direction, and timing with respect to tides.

Chapter 4 ("High frequency trends in the isotopic composition of Superstorm Sandy" by Stephen Good et al.) introduces a novel methodology for analyzing storm dynamics. By analyzing stable isotopes in rainwater samples collected by citizen scientists, the authors gain insight into the mixing of air masses that create storms such as Sandy (Good et al., 2014). Furthermore, as efforts intensify to extend the record of storm impacts into the past beyond the historical record, studies such as this one will prove critical. Accurate interpretation of speleothem isotopic data, for example, depends on understanding the isotopic signature of rainfall events as well as the effects of percolation through regolith and other in-situ factors (Evans et al., 2013; Frappier, 2013).

Chapters 5 ("Hurricane Sandy: Did Beach Nourishment Save New Jersey?" by Adam Griffith and others) and 6 ("Observations of the Influence of Regional Beach Dynamics on the Impacts of Storm Waves on the Connecticut Coast During Hurricanes Irene and Sandy" by James Tait and Ezgi Akpinar Ferrand) examine the effects of storm surge and wave action on shoreline structures in New Jersey and Connecticut. Both studies suggest that a greater distance between structures and the mean high water line results in significantly less damage to the structures. In addition, nourishment of beaches did seem to reduce damage to shoreline structures in New Jersey, but only slightly.

Chapters 7 ("Recognizing Past Storm Events in Sediment Cores Based On Comparison To Recent Overwash Sediments Deposited By Superstorm Sandy" by J. Bret Bennington and E. Christa Farmer) and 8 ("Trace Metals as a Tool for Chronostratigraphy in Sediment Cores from South Shore Barrier Beach Marshes in Long Island, NY" by Christa Farmer and Bret Bennington) utilize the washover sediments generated by Sandy to better interpret the geological record of storm impacts, a critical tool in paleotempestology for extending the temporal record of storms beyond the records available from satellites and modern instruments.

The New York metropolitan area (encompassing the urban region of New York City, northern New Jersey, and western Connecticut) is currently the eighth largest in the world with a combined population of approximately 20.66 million residents (Demographia, 2014), many of whom live at or near sea level within a few kilometers of the coast. This region is highly vulnerable to Atlantic storms and will become even more so as sea level continues to rise and ocean temperatures continue to warm over the next century (IPCC, 2013). Recent studies suggest that the strongest storms may become more frequent (Gleixner et al., 2014; Holland and Bruyere, 2014; Grinsted et al., 2013). Superstorm Sandy, for all of the destruction and misery that the storm inflicted on the region, was a timely wake-up call and harbinger of things to come. It is critical that we learn all that can be learned from Sandy's impact on the New York metropolitan region so that we can better understand the history of major storms in this area and so that we can better prepare to weather the forces of future storms.

ACKNOWLEDGMENTS

The research reported on in this volume was originally presented at a Geological Society of America theme session titled, "Learning from the Impacts of Superstorm Sandy" on October 27, 2013. We thank all of our colleagues in that session for participating and for making the session so timely and informative. We also thank Steve Leone, one of our former Hofstra undergraduate students, for coming up with the idea for proposing a Superstorm Sandy session and for doing much of the work organizing and running it. Steve was involved in the field work collecting cores from Sandy overwash lobes and his enthusiasm for the project and gentle cajoling were instrumental in making the session out of which this book grew a reality. We are also grateful to Candice Janco, Sean Coombs, and Marisa LaFleur at Elsevier for encouraging us to share the research reported on in the GSA theme session by pursuing this book project, and for their patience in working with us to see the project to completion. This whole project would not have been possible without all kinds of support from many of our colleagues at Hofstra University: special thanks go to Hofstra College of Liberal Arts and Sciences Dean Bernard Firestone for his assistance in securing funding to support our research, travel, writing, and editing.

REFERENCES

Blake, Eric S., Kimberlain, Todd B., Berg, Robert J., Cangialosi, John P., Beven II, John L. National Hurricane Center (February 12, 2013) (PDF). Hurricane Sandy: October 22–29, 2012 (Tropical Cyclone Report). United States National Oceanic and Atmospheric Administration's National Weather Service (accessed 05.08.14).

Demographia, 2014. Demographia World Urban Areas, 10th Ed. May 2014 Revision (PDF). http://www.demographia.com/db-worldua.pdf.

Evans, M.N., Tolwinski-Ward, S.E., Thompson, D.M., Anchukaitis, K.J., 2013. Applications of proxy system modeling in high resolution paleoclimatology. Quaternary Science Reviews 76, 16–28.

Frappier, A.B., 2013. Masking of interannual climate proxy signals by residual tropical cyclone rainwater: evidence and challenges for low-latitude speleothem paleoclimatology. Geochemistry Geophysics Geosystems 14, 3632–3647.

Good, S.P., Mallia, D.V., Lin, J.C., Bowen, G.J., 2014. Stable isotope analysis of precipitation samples obtained via crowdsourcing reveals the spatiotemporal evolution of Superstorm Sandy. Plos One 9.

Gleixner, S., Keenlyside, N., Hodges, K.I., Tseng, W.L., Bengtsson, L., 2014. An inter-hemispheric comparison of the tropical storm response to global warming. Climate Dynamics 42, 2147–2157.

Grinsted, A., Moore, J.C., Jevrejeva, S., 2013. Projected Atlantic hurricane surge threat from rising temperatures. Proceedings of the National Academy of Sciences of the United States of America 110, 5369–5373.

Holland, G., Bruyere, C.L., 2014. Recent intense hurricane response to global climate change. Climate Dynamics 42, 617–627.

IPCC, 2013. Climate Change 2013 the physical science basis. Contribution of Working Group I to the Fifth Assessment Report of the Intergovernmental Panel on Climate Change [Stocker, T.F., D. Qin, G.-K. Plattner, M. Tignor, S.K. Allen, J. Boschung, A. Nauels, Y. Xia, V. Bex and P.M. Midgley (eds.)]. Cambridge University Press, Cambridge, United Kingdom and New York, NY, USA; 1; 1535 pp.

Lixion, A. Avila, John Cangialosi (December 14, 2011). "Hurricane Irene Tropical Cyclone Report" (PDF). National Hurricane Center (accessed 05.08.14)

Rappaport, E.N., 2014. Fatalities in the United States from Atlantic Tropical Cyclones. Bulletin of the American Meteorological Society 95, 341–346.

CHAPTER 2

Measuring Storm Tide and High-water Marks Caused by Hurricane Sandy in New York

A.E. Simonson and R. Behrens

ABSTRACT

In response to Hurricane Sandy, personnel from the U.S. Geological Survey (USGS) deployed a temporary network of storm-tide sensors from Virginia to Maine. During the storm, real-time water levels were available from tide gages and rapid-deployment gages (RDGs). After the storm, USGS scientists retrieved the storm-tide sensors and RDGs and surveyed high-water marks. These data demonstrate that the timing of peak storm surge relative to astronomical tide was extremely important in southeastern New York. For example, along the south shores of New York City and western Suffolk County, the peak storm surge of 6–9 ft generally coincided with the astronomical high tide, which resulted in substantial coastal flooding. In the Peconic Estuary and northern Nassau County, however, the peak storm surge of 9 ft and nearly 12 ft, respectively, nearly coincided with normal low tide, which helped spare these communities from more severe coastal flooding.

Keywords: storm surge; storm tide; high-water marks; Hurricane Sandy; coastal flooding; New York emergency management

2.1 INTRODUCTION

On the evening of October 30, 2012 at approximately 1:00 am GMT Post-Tropical Cyclone Hurricane Sandy made landfall 5 miles southwest of Atlantic City, New Jersey (National Hurricane Center, 2012), producing Category 1 hurricane force winds and coastal flooding in the Northeast region of the United States (Fanelli et al., 2013). Prior to landfall, the USGS deployed a temporary monitoring network of

Learning from the Impacts of Superstorm Sandy. http://dx.doi.org/10.1016/B978-0-12-801520-9.00002-X

water-level and barometric-pressure sensors at 224 locations along the Atlantic coast from Virginia to Maine, which recorded the timing, areal extent, and magnitude of storm tide and coastal flooding (McCallum et al., 2013).

As defined by the National Oceanic and Atmospheric Administration (NOAA), storm tide is the maximum water level elevation during storm events, and can be a combination of storm surge, astronomical tide, and river runoff (Fanelli et al., 2013). Storm surge is the onshore rush of seawater caused by the high wind and the low-pressure center of a hurricane or other storm, and the magnitude of storm surge is dependent upon the orientation of the coastline with the storm track, the intensity of the storm, and the local bathymetry (Fanelli et al., 2013).

After the storm, more than 950 HWMs were flagged with an emphasis on New Jersey and New York, of which 653 marks were surveyed to establish elevation (McCallum et al., 2013). These efforts were undertaken as part of a larger coordinated Federal emergency response as outlined by the Stafford Act under a directed mission assignment by the Federal Emergency Management Agency (FEMA). The USGS has conducted similar deployments and other studies to examine the effects of hurricanes and tropical storms and to understand potential impacts on coastal communities and habitats; details are available online at http://coastal.er.usgs.gov/hurricanes.

2.2 BACKGROUND: HURRICANE SANDY STORM-TIDE MONITORING

2.2.1 Before the Storm: Long-term Network

The low-lying, highly-populated coastal areas of southeastern New York are vulnerable to tidal flooding, and emergency managers need adequate flood warnings during hurricanes and other coastal storms. To address this need for instantaneous information on coastal flooding, the USGS has operated a network of real-time tidal water-elevation stations since 1997 in southeastern New York, with the cooperative support of state and local municipalities. Each tidal water-elevation station, or tide gage, is equipped with a pressure sensor connected to a data-collection platform inside a weather-resilient shelter. Data are recorded at 6-minute intervals, transmitted via satellite every hour, and uploaded to the

USGS National Water Information System (NWIS). When a station detects water levels above the National Weather Service (NWS) minor coastal-flood elevation threshold, it increases the frequency of satellite transmissions to 6-minute intervals. Real-time tidal water-elevation data at New York sites are available at http://waterdata.usgs.gov/ny/nwis/current/?type=tidal. During Hurricane Sandy, there were 13 long-term tide gages in southeastern New York, 10 of which were equipped with telemetry. Data from the USGS permanent monitoring sites were used to compute the residual water level (storm surge, when positive) by subtracting the predicted astronomical tide level from the observed water level (Fanelli et al., 2013). The water-level elevations and storm-surge calculations were available to emergency managers and residents during the storm.

In addition to the coastal network, there is a long-term network of more than 200 stream gages in New York, which is part of a nation-wide network of more than 9000 gages (Lurry, 2011). During the hurricane, there were six real-time stream gages in southeastern New York that recorded storm tide. These sites measured water levels at 15-min intervals and transmitted data hourly to the NWIS website: http://waterdata.usgs.gov/ny/nwis/current/?type=flow. Because many Long Island stream gages have data records extending back to the 1930s (Wells, 1960), they provide valuable historical context to this and other recent storms.

2.2.2 Before the Storm: Short-term Network

From October 26 to 28, 2012, the NY USGS deployed 39 storm-tide sensors, 4 wave-height sensors, 11 barometric-pressure sensors, and 4 RDGs throughout Long Island, New York City, and Westchester County as part of a larger, coordinated multicenter effort. Sensor locations were selected to supplement the existing USGS long-term tide gage network and to ensure that sufficient data were collected in areas where NOAA models predicted significant coastal flooding. Storm-tide and wave-height sensors are non-vented pressure transducers programmed to record at 30-s and 2-s intervals, respectively. Both the storm-tide and wave-height sensors are unvented, so barometric-pressure sensors are installed nearby to compensate for the pressure change during the storm. Pressure transducers were installed in aluminum or PVC housings and attached to stable structures, such as bulkheads and pilings.

Sensors were deployed at low tide, preferably submerged below the water surface to collect complete tidal cycles. RDGs equipped with sensors to record water level and meteorological data were usually attached to bridge railings. The RDG data were transmitted with telemetry to the NWIS website during the storm.

Further details of this deployment are available in McCallum et al. (2013) and in the online mapper at: http://water.usgs.gov/floods/events/2012/sandy/sandymapper.html. Additional storm-surge analysis of short-term storm-tide sensors is available in Behrens (2013).

2.2.3 After the Storm: High-water Marks and Surveying

Within 10 days of landfall, the short-term water-level sensors were collected, surveyed to the North American Vertical Datum of 1988 (NAVD 88) and processed following standard USGS protocols (McGee et al., 2005; McCallum et al., 2012), which included correcting water pressure for changes in barometric pressure and salinity, and comparing recorded elevations to nearby independent high-water marks and long-term gages. In southeastern New York alone, more than 350 high-water marks were flagged and surveyed. High-water marks – such as debris, seed lines, and mud lines – were documented as verification of the storm-tide sensor data and as indicators of peak storm tide. Because precipitation and clean-up efforts can easily destroy such marks, they must be identified as soon as possible after an event. To determine the elevation of the HWMs and short-term network sensors, a combination of differential leveling and survey-grade Global Navigation Satellite Systems (GNSS) equipment was used to determine the elevation above NAVD 88 (Rydlund and Densmore, 2012). For quality assurance, National Geodetic Survey (NGS) benchmarks with known elevations were surveyed throughout the study area for vertical control (see Table 2 in McCallum et al., 2013).

2.3 RESULTS

This paper describes a subset of the overall USGS effort, and focuses on four representative geographic areas and their adjacent water bodies: (1) southern Suffolk County along the shore of Great South Bay; (2) New York City along the shore of Lower New York Bay/Raritan Bay; (3) eastern Suffolk County along the shore of the Peconic Estuary; and

(4) northern Nassau County along the shore of western Long Island Sound (see Figure 2.1). The data shown in Table 2.1 were chosen to demonstrate the range in storm-tide elevations and the fidelity of the record for nearby sites and were rounded to 0.1 ft for simplicity. Our storm-tide data are from Tables 3, 4, and 6 of McCallum et al. (2013), which is available online at http://pubs.usgs.gov/of/2013/1043, and storm-surge data are from the NWIS tide gage graphics recorded during the hurricane (USGS, 2012a–d). Peak storm-tide data are affected by wave action to various degrees, depending on the measurement techniques and site conditions under which the storm-tide elevations were collected (C. Schubert, U.S. Geological Survey, written communications, 2014).

In southern Suffolk County along the shore of Great South Bay, data were collected at two real-time stations during Hurricane Sandy, a long-term USGS tide gage (01309225, Great South Bay at Lindenhurst, NY) and an RDG (403836073154775, State Boat Channel at Captree Island, NY). The RDG recorded a peak storm-tide elevation of 5.2 ft and the tide

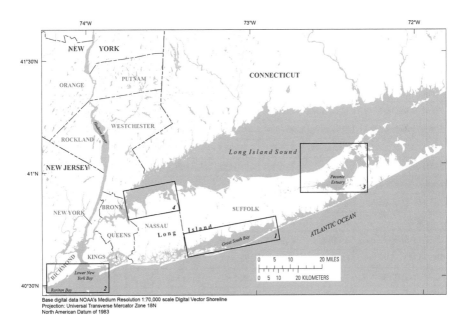

Fig. 2.1. Map showing location of study areas along their adjacent water bodies: (1) southern Suffolk County along the shore of Great South Bay; (2) New York City along the shore of Lower New York Bay/Raritan Bay; (3) eastern Suffolk County along the shore of the Peconic Estuary; and (4) northern Nassau County along the shore of western Long Island Sound.

Table 2.1 Select Peak Storm-tide Data from Hurricane Sandy for U.S. Geological Survey Permanent Monitoring Sites, Temporarily Deployed Sites, and High-water Mark Data, Modified from McCallum et al. (2013)

Site Identification	County	Latitude North, Decimal Degrees	Longitude West, Decimal Degrees	Site Type	Type of Data Recorded	Peak Storm Tide, Feet NAVD 88
Southern Suffolk County, Great South Bay						
HWM-NY-SUF-403	Suffolk	40.6853	−73.2799	High-water mark	Storm tide	7.4
SSS-NY-SUF-022WL	Suffolk	40.6852	−73.2799	Water level	Storm tide	6.8
SSS-NY-SUF-021WL	Suffolk	40.7492	−73.0134	Water level	Storm tide	6.7
HWM-NY-SUF-622	Suffolk	40.6783	−73.3330	High-water mark	Storm tide	6.6
01309225	Suffolk	40.6693	−73.3557	Real-time tide gage	Storm tide	6.5
SSS-NY-SUF-027WL	Suffolk	40.7476	−73.1504	Water level	Storm tide	6.1
HWM-NY-SUF-405	Suffolk	40.6912	−73.2772	High-water mark	Storm tide	5.8
SSS-NY-SUF-019WL	Suffolk	40.6593	−73.2649	Water level	Storm tide	5.6
403836073154775	Suffolk	40.6433	−73.2631	Real-time RDG	Storm tide	5.2
SSS-NY-SUF-018WL	Suffolk	40.6347	−73.2022	Water level	Storm tide	4.1
New York City, Raritan Bay						
SSS-NY-RIC-003WL	Richmond	40.5019	−74.2303	Water level	Storm tide	16.0
SSS-NY-RIC-001WV	Richmond	40.5939	−74.0598	Wave height	Wave height	15.1
SSS-NY-RIC-001WL	Richmond	40.5939	−74.0598	Water level	Storm tide	15.0
HWM-NY-RIC-982	Richmond	40.5458	−74.1238	High-water mark	Storm tide	14.0
SSS-NY-KIN-001WL	Kings	40.5800	−74.0116	Water level	Storm tide	13.3
SSS-NY-RIC-004WL	Richmond	40.5435	−74.1277	Water level	Storm tide	13.2
HWM-NY-RIC-704	Richmond	40.5024	−74.2311	High-water mark	Storm tide	13.2
HWM-NY-RIC-719	Richmond	40.5939	−74.0683	High-water mark	Storm tide	12.7
01311875	Kings	40.5737	−73.8851	Real-time tide gage	Storm tide	10.7
Eastern Suffolk County, Peconic Estuary						
SSS-NY-SUF-005WL	Suffolk	40.9161	−72.6377	Water level	Storm tide	7.9
01304562	Suffolk	40.9178	−72.6387	Real-time tide gage	Storm tide	7.7
SSS-NY-SUF-014WL	Suffolk	40.9907	−72.4707	Water level	Storm tide	7.4
HWM-NY-SUF-307	Suffolk	40.9898	−72.4708	High-water mark	Storm tide	7.1
SSS-NY-SUF-008WL	Suffolk	40.8933	−72.5030	Water level	Storm tide	6.5
01304200	Suffolk	41.1366	−72.3068	Real-time tide gage	Storm tide	6.4
SSS-NY-SUF-015WL	Suffolk	41.1010	−72.3614	Water level	Storm tide	6.4
SSS-NY-SUF-009WL	Suffolk	41.0020	−72.2903	Water level	Storm tide	6.3
HWM-NY-SUF-436	Suffolk	40.8935	−72.5033	High-water mark	Storm tide	6.3
SSS-NY-SUF-024WL	Suffolk	41.0732	−71.9344	Water level	Storm tide	6.1

Table 2.1 Select Peak Storm-tide Data from Hurricane Sandy for U.S. Geological Survey Permanent Monitoring Sites, Temporarily Deployed Sites, and High-water Mark Data, Modified from McCallum et al. (2013) *(cont.)*

Site Identification	County	Latitude North, Decimal Degrees	Longitude West, Decimal Degrees	Site Type	Type of Data Recorded	Peak Storm Tide, Feet NAVD 88
Northern Nassau County, Long Island Sound						
HWM-NY-NAS-518	Nassau	40.8350	−73.7287	High-water mark	Storm tide	10.8
01302250	Nassau	40.8662	−73.7102	Real-time tide gage	Storm tide	10.3
SSS-NY-NAS-008WL	Nassau	40.8662	−73.7102	Water level	Storm tide	10.3
HWM-NY-NAS-700	Nassau	40.8875	−73.5636	High-water mark	Storm tide	10.2
SSS-NY-NAS-006WL	Nassau	40.8875	−73.5636	Water level	Storm tide	10.2
SSS-NY-NAS-001WL	Nassau	40.8779	−73.5306	Water level	Storm tide	10.1
01302845	Nassau	40.9051	−73.5932	Real-time tide gage	Storm tide	10.1
HWM-NY-NAS-938	Nassau	40.8915	−73.6357	High-water mark	Storm tide	10.0
SSS-NY-NAS-007WL	Nassau	40.8572	−73.4633	Water level	Storm tide	10.0
01302600	Nassau	40.8886	−73.6380	Real-time tide gage	Storm tide	9.9

gage recorded 6.5 ft NAVD 88, with surrounding storm-tide data in the Great South Bay ranging from 4.1 to 7.4 ft (Table 2.1). The graph of data from the aforementioned long-term tide gage showed a calculated storm surge of approximately 6.4 ft (Figure 2.2a), which generally coincided with astronomical high tide in the area, resulting in widespread coastal flooding.

In New York City, along the shore of Lower New York Bay/Raritan Bay, storm-tide sensors and long-term tide gages in Richmond and Kings Counties recorded peak storm-tide elevations of 10.7–16.0 ft NAVD 88 (Table 2.1). Some of the variation in peak storm tide is related to wave action. For example, Figure 2.3 shows the data variability of the wave-height sensor near the Verrazano Bridge (SSS-NY-RIC-001WV) relative to data from the water-level sensor that was sheltered from wave action in Great Kills Harbor (SSS-NY-RIC-004WL). As shown in Figure 2.2b, the nearby USGS tide gage at Rockaway Inlet (01311875) indicated a peak storm surge of approximately 9.1 ft, which generally coincided with astronomical high tide and resulted in extensive coastal flooding in the area.

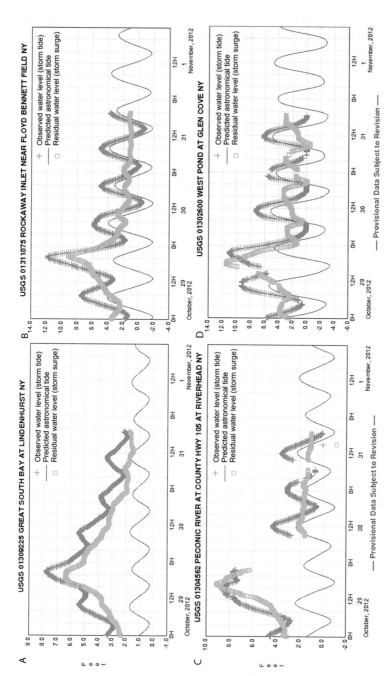

Fig. 2.2. Hydrographs for October 29–November 1, 2012, at (A) 01309225 Great South Bay at Lindenhurst, (B) 01311875 Rockaway Inlet near Floyd Bennett Field, (C) 01304562 Peconic River at County Highway 105 at Riverhead, and (D) 01302600 West Pond at Glen Cove. Residual water level (green squares) is shown in feet and is calculated from difference between observed water elevation and predicted (astronomical) tide elevation. Water elevation record (orange crosses) is shown in feet above NGVD 1929 (National Geodetic Vertical Datum of 1929). Predicted astronomical tide (blue line) is shown in feet above NGVD 1929 for nearby National Ocean Service (NOS) tidal-prediction stations, (A) NOS 1255; (B) NOS 1281; (C) NOS 1209; (D) NOS 1165. Time is shown in Eastern Standard Time. Data are provisional and subject to revision. Figures modified from USGS (2012a–d).

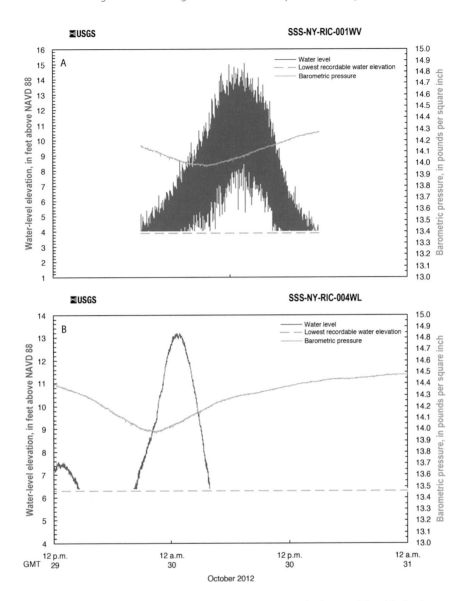

Fig. 2.3. *Hydrographs for October 30, 2012 for SSS-NY-RIC-001WV wave-height sensor (A) and for October 29–31, 2012 for SSS-NY-RIC-004WL water-level sensor (B). Water level elevation (blue line) is shown in feet above NAVD 1988. Barometric pressure (yellow line) is shown in pounds per square inch. Time is shown in Eastern Standard Time. Data are provisional and subject to revision. Figures are available from the online mapper at: http://water.usgs.gov/floods/events/2012/sandy/sandymapper.html.*

In contrast, eastern Suffolk County along the shore of the Peconic Estuary experienced less coastal flooding because maximum surge nearly coincided with astronomical low tide. Peak storm-tide elevation in this area ranged from 6.1 to 7.9 ft NAVD 88. The USGS real-time tide gage at Riverhead (01304562) and a nearby storm-tide sensor (SSS-NY-SUF-005WL) recorded maximum water-level elevations of 7.7 and 7.9 ft, respectively (Table 2.1), and the peak storm surge calculated at the Riverhead tide gage was approximately 9.2 ft (Figure 2.2c).

The northern shore of Nassau County along western Long Island Sound also experienced widespread coastal flooding. Overall, storm-tide data in this area ranged from 9.9 to 10.8 ft NAVD 88 (Table 2.1). The peak storm surge calculated in the area was approximately 11.5 ft (Figure 2.2d) at the West Pond tide gage (01302600), which nearly coincided with astronomical low tide, sparing surrounding communities from much greater flooding. For historical context, the Mill Neck stream gage (0130300) was established in 1937 and its highest recorded water level occurred during the unnamed 1938 hurricane (Busciolano et al., 2009). Although this permanent monitoring site was discontinued in 2011, both the temporary sensor (SSS-NY-NAS-006WL) and high-water mark (HWM-NY-NAS-700) from inside of the gage indicated water levels of 10.2 ft NAVD 88, only 0.06 ft less than the historical 1938 peak water-level elevation (C. Schubert, U.S. Geological Survey, written communications, 2014), which emphasizes both the historical nature of this storm and what could have happened if peak storm surge had occurred during high tide.

A summary of peak storm-tide data are shown for southeastern New York in Figure 2.4, which is derived from the permanent network of tide and stream gages, the short-term network of RDGs, storm-tide and wave-height sensors, and the post-storm collection of high-water marks. These data are also available in the interactive USGS online mapper at: http://water.usgs.gov/floods/events/2012/sandy/sandymapper.html.   As evidenced by the red and pink circles, the western part of the study area – Richmond, Kings, Queens, and Nassau counties – experienced storm tide from about 8 to 19.5 ft NAVD 88. The yellow and blue circles, however, indicate that most of southern and eastern Suffolk County experienced storm tide of about 4 to 8 ft NAVD 88. For all peak storm-tide data, see McCallum et al. (2013).

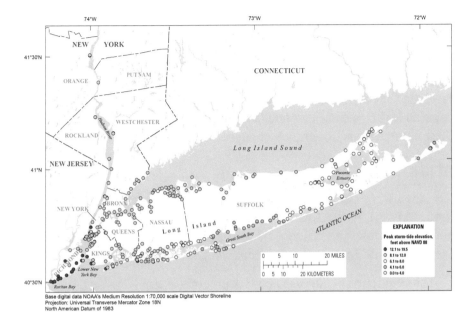

Base digital data NOAA's Medium Resolution 1:70,000 scale Digital Vector Shoreline
Projection: Universal Transverse Mercator Zone 18N
North American Datum of 1983

Fig. 2.4. Map showing peak storm-tide elevations produced by Hurricane Sandy in New York, in feet above NAVD 88. These data are also available from the USGS interactive storm-tide mapper at http://water.usgs.gov/floods/ events/2012/sandy/sandymapper.html. Data are provisional and subject to revision.

2.4 SUMMARY AND CONCLUSIONS

In the days before Hurricane Sandy, the USGS installed numerous sensors along the northeastern coast of the United States to measure storm-tide elevation and provide the timing, water depth, and duration of storm-tide flooding. During Hurricane Sandy's approach and landfall, local emergency managers and NOAA forecasters were able to observe water levels in real time from the existing USGS network of tide gages and from the temporary RDGs installed in advance of the storm. As shown in Figure 2.4, the peak water levels recorded by the storm-tide sensors and surveyed HWMs ranged from 4 to 10 ft (above NAVD 88) in southern Suffolk County, from 6 to 8 ft in the Peconic Estuary of eastern Suffolk County, from 8 to 12 ft in northern Nassau County, and from 9 to 17 ft in New York City.

Data from permanent monitoring sites in the USGS network were used to compute the storm surge magnitude. The timing of peak storm surge relative to astronomical tide was extremely important. For example, along the south shores of New York City and western Suffolk County, the peak storm surge of 6–9 ft generally coincided with the astronomical high tide (Figure 2.2a and b), which resulted in widespread coastal flooding in these areas. In the Peconic Estuary and northern Nassau County, however, the peak storm surge of 9 ft and nearly 12 ft, respectively, nearly coincided with normal low tide (Figure 2.2c and d), which helped spare these communities from much greater coastal flooding.

ACKNOWLEDGMENTS

The long-term network of tide gages in southeastern New York is funded in part by the Town of Hempstead Department of Conservation and Waterways, the Village of Freeport, and the New York State Department of Environmental Conservation.

The long-term stream gage network in southeastern New York is funded in part by Suffolk County Water Authority, Suffolk County Department of Health, New York City Department of Environmental Protection, and New York State Department of Environmental Conservation.

The storm-tide sensor and high-water mark efforts were undertaken as part of a larger coordinated Federal emergency response as outlined by the Stafford Act under a directed mission assignment by the Federal Emergency Management Agency.

The authors would also like to thank J. Finkelstein, R. Busciolano, C. Schubert, and B. McCallum for their helpful comments.

REFERENCES

Behrens, R., 2013. Historical Storm Surges on Long Island During Extreme Weather Events. Stony Brook University, Stony Brook, Research Project, pp. 44.

Busciolano, R., Lange, A., Simonson, A., 2009. Water Resources Data, New York, Water Year 2009 [Online]. U.S. Geological Survey Water-Data Report NY-09-02. Available at: http://wdr.water.usgs.gov/wy2009/pdfs/01303000.2009.pdf (accessed 01.08.2014).

Fanelli, C., Fanelli, P., Wolcott, D., 2013. NOAA water level & meteorological data report – Hurricane Sandy. U.S. Department of Commerce, National Oceanic and Atmospheric Administration, National Ocean Service Center for Operational Oceanographic Products and Services, pp. 62

Lurry, D.L., 2011. How does a U.S. Geological Survey streamgage work? U.S. Geological Survey Fact Sheet 3001, 2.

McCallum, B., Painter, J., Frantz, E., 2012. Monitoring inland storm tide and flooding from Hurricane Irene along the Atlantic Coast of the United States. U. S. Geological Survey Open-File Report 1022, 28.

McCallum, B., Wicklein, S., Reiser, R., Busciolano, R., Morrison, J., Verdi, R., Painter, J., Frantz, E., Gotvald, A., 2013. Monitoring storm tide and flooding from Hurricane Sandy along the Atlantic coast of the United States, October 2012. U. S. Geological Survey Open-File Report 1043, 42.

McGee, B.D., Goree, B.B., Tollett, R.W., Woodward, B.K., and Kress, W.H., 2005. Hurricane Rita surge data, southwestern Louisiana and southeastern Texas, September to November 2005 [Online]. U.S. Geological Survey Data Series 220. Available at: http://pubs.usgs.gov/ds/2006/220/ (accessed 01.08.2014).

National Hurricane Center, 2012. Post-tropical cyclone sandy update: national oceanic and atmospheric administration [Online]. National Weather Service, Available at: http://www.nhc.noaa.gov/archive/2012/al18/al182012.update.10300002.shtml (accessed 01.08.2014).

Rydlund, P.H., Jr., and Densmore, B.K., 2012. Methods of practice and guidelines for using survey-grade global navigation satellite systems (GNSS) to establish vertical datum in the United States Geological Survey. U.S. Geological Survey Techniques and Methods, Book 11, Chapter D1, pp. 102.

U.S. Geological Survey, 2012a. National Water Information System data available on the World Wide Web (Water Data for the Nation). Available at: http://waterdata.usgs.gov/ny/nwis/uv/?site_no = 01309225.[Accessed October 31, 2012].

U.S. Geological Survey, 2012b. National Water Information System data available on the World Wide Web (Water Data for the Nation). Available at: http://waterdata.usgs.gov/ny/nwis/uv/?site_no = 01311875 (accessed 31.10.2012).

U.S. Geological Survey, 2012c. National Water Information System data available on the World Wide Web (Water Data for the Nation). Available at: http://waterdata.usgs.gov/ny/nwis/uv/?site_no = 01302600 (accessed 31.10.2012).

U.S. Geological Survey, 2012d. National Water Information System data available on the World Wide Web (Water Data for the Nation). Available at: http://waterdata.usgs.gov/ny/nwis/uv/?site_no = 01304562.(accessed 31.10.2012).

Wells, J., 1960. Compilation of records of surface waters of the United States through September 1950, Part 1-B, North Atlantic slope basins New York to York River. U. S. Geological Survey Water Supply Paper 1302, 679.

CHAPTER 3

Superstorm Sandy and Staten Island: Learning from the Past, Preparing for the Future

Alan I. Benimoff, William J. Fritz and Michael Kress

ABSTRACT

We have utilized a GIS (Geographic Information System) in order to study the massive property loss and loss of life from Superstorm Sandy. Using an Advanced Circulation Model for Oceanic, Coastal, and Estuarine waters (ADCIRC) developed at the Renaissance Computing Institute (RENCI) at the University of North Carolina at Chapel Hill, we modeled the storm surge. In many key locations on the southeastern shore of Staten Island, agreement was within 0.02 meters of the high-water mark recorded by the United States Geological Survey. A further result of this study is a five-point plan: (1) Protect the existing natural barriers – the beaches and dunes; (2) build them higher; (3) rezone in the flood zone and buy up as many properties as possible in low-lying areas, turning them into parkland; (4) be very careful about engineering solutions such as sea barriers because they will not only be expensive but necessarily protect one area at the expense of another; (5) provide education to residents of Staten Island.

Keywords: Superstorm Sandy; Staten Island; Geographic Information System; advanced circulation model for oceanic coastal and estuarine waters (ADCIRC); storm surge

3.1 INTRODUCTION

On October 29, 2012, Superstorm Sandy inflicted a devastating blow to our portion of the East Coast. The storm provided a wake-up call to the entire New York Metropolitan area but especially to Staten Island, The Rockaways, Breezy Point, and other low-lying coastal areas. As we move into the future and face other potentially catastrophic weather

Learning from the Impacts of Superstorm Sandy. http://dx.doi.org/10.1016/B978-0-12-801520-9.00003-1

events, we must use what we have learned from past storms such as Sandy to serve as a road map for the future.

In addition, it is critical to embrace the importance of taking an interdisciplinary approach toward formulating a response to and preparing for natural disasters (Benimoff et al., 2013a,b,c). As we look back at the history of weather disasters, scientists have made predictions, but few policymakers and members of the public have acted on their warnings. In order for these warnings to be received, the scientific community must bring into their discussions a variety of people such as scientists, geologists, engineers, social scientists, counselors, political scientists, politicians, economists, community members, city planners, emergency responders, government agencies, and the business community. This will provide a significantly more effective team approach in the face of a likely increase in the frequency and severity of tropical storms.

3.1.1 An Active, But Overlooked Storm History

We recently held an interdisciplinary forum (http://www.csi.cuny.edu/sandyforum/news.html) in which we brought together community experts to deal with a number of aspects of storm surge and flooding. In addition to the geologic issues, we also dealt with topics such as the human impact, the economic and political aspects, and the need for more education.

Severe storms (extratropical and hurricane) experienced in the New York Harbor Area are plotted in the time line shown in Figure 3.1. At least 31 of these severe storms have occurred in the past 379 years for an average of one storm every 12 years. It is unclear if the storms, post-1900, represent an increase in frequency or simply the lack of historical data in the seventeenth and eighteenth centuries. We suggest that the true average of storm surges in the New York area may be significantly less than 12 years.

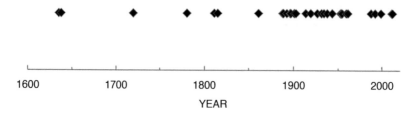

Fig. 3.1. Hurricane and major storm events affecting the NYC area. (Source pre-1964: U.S. Army Corps of Engineers, 1964)

In 1932, a hurricane of unknown strength with a 4+ m surge (based on our analysis of newspaper photos) struck the area, and in 1938, an unnamed Category 3 hurricane, sometimes called the "Long Island Express", produced a surge of approximately 6 m. These storms, for the most part, went unnoticed, at least on Staten Island, because the surges flowed across undeveloped marshland.

The summer storm of 1988 caused severe flooding in the Sweetbrook Drainage Basin (Benimoff, 2012). The winter storm of 1992 resulted in extensive flooding from storm surge. Severe flooding occurred during Tropical Storm Floyd (1999) when 154.43 mm of rain fell, as recorded at the CSI weather station. The hydrologic response to Hurricane Irene (2011) was extensive flooding in the Richmond town area, where the gauging station (USGS 01376534) at the Richmond Creek Bluebelt BMP3 recorded a stage of 1.54 m instead of the normal stage of 0.3048 m; extensive flooding in the Willowbrook area; and extensive flooding in the South, Midland, and Fox Beach areas. The Bluebelt is a storm water management system designed and implemented by the New York City Department of Environmental Protection that utilizes existing streams by installing Best Management Practices (BMPs).

A view of past Atlantic hurricane tracks (Scileppi and Donnelly, 2007; Elsener et al., 1999) shows that most Atlantic hurricanes start off the west coast of Africa traveling west and are deflected parallel up the east coast of the U.S. Superstorm Sandy was nearly a "coast normal" hurricane, as it struck the New Jersey coast while traveling at an angle almost perpendicular to the coastline. This was due to several factors, one of which was a "negatively tilted trough" (Kieffer, 2013; Blake et al., 2013) in the polar jet stream that was in close proximity to the subtropical jet stream. There was also a blocking high in the north that prevented Sandy from going out to sea. The track of Superstorm Sandy is an exceedingly rare event as calculated by Hall and Sobel (2013) with a return period of 714 years. They did not factor in the storm surge because they indicate that storm surge depends on many factors. This track put the northeastern quadrant of the hurricane in the New York Metropolitan area. In that quadrant, the forward speed and the wind speed are additive, producing the greatest storm surge. Couple this with the right angle in the coastline (Figure 3.2) and it is clear that Staten Island was in particular danger from Superstorm Sandy.

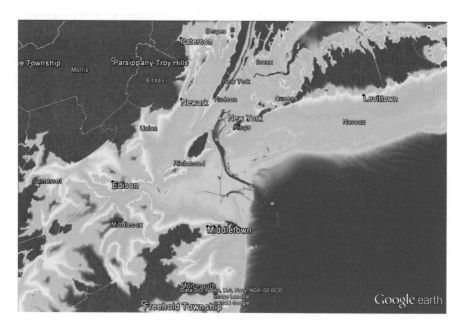

Fig. 3.2. DEM of NYC region showing right angle in coastline. Color scale goes from red (40 m above sea level and higher) to blue (20 m below sea level or lower). Map generated by the CUNY Interdisciplinary High-Performance Computing Center at the College of Staten Island. Base Image: 2013 Terrametrics; U.S. Navy NGA GEBCO; 2013 Google Earth.

3.2 STORM DYNAMICS, NYC'S UNFORTUNATE LOCATION, AND A DANGEROUSLY SHIFTING LANDSCAPE

Every storm is different, as strength, eye track, tides, and other weather systems are all variable factors. The surge begins as a low-pressure bulge in the ocean in the eye of the storm (similar to water rising into a vacuum cleaner). Winds pile water on top of the bulge and the tide then lifts the water to an even higher level. The New York metropolitan area is located in a particularly vulnerable area. In fact, it is the most dangerous place on the Eastern Seaboard. The right angle of the coastline (Figure 3.2), a result of the intersection of the New Jersey Shore and Long Island, and counterclockwise rotation, pushes wind and water against Staten Island and up New York Harbor. This surge water is then trapped by winds forcing water westward along Long Island Sound.

Staten Island is particularly vulnerable, in this regard. The narrowing passage created by the Long Island and New Jersey shores, combined with a sea floor ramp that grows increasingly shallow, pressurizes the water and aims it against the South Shore, significantly increasing the height and intensity of the surge. Twenty-three Staten Island residents died as a result of Sandy.

Increasing sea levels only exacerbate the surge, and higher sea levels should now be considered as the new normal. For the past 500 years, the sea level has been increasing at about a foot per century, and that rate is likely to increase perhaps to as much as two to five feet per century (Rahmstorf, 2012). These effects have been hidden because the shore has been developing faster than the sea-level rise growth which is not sustainable, and extremely dangerous.

The 1902 USGS topographic map of Staten Island, NY (Figure 3.3) shows extensive tidal wetlands on its eastern shore. During this time,

Fig. 3.3. 1902 topographic map of Staten Island, New York (from USGS 1902 Topographic Map). Note marshland and tidal channels that were at or below sea level in 1902 along the East Shore of Staten Island when sea level was about a foot lower than today (2014).

these tidal wetlands were able to absorb the waters from hurricane surges. Barrier islands, marshes, coastal dune fields, estuaries, and bays serve as nature's sponges, absorbing the force of a storm surge and storing water, mitigating damage and flooding. As a result of development, we have hardscaped these sponges, making these areas extremely susceptible to flood damage.

Benimoff (2010), in his GIS study, reported on extensive urbaniza-tion of the SLOSH (Sea Lake and Overland Surges from Hurricanes) zones during the period of 1898 through 2008. Using pre-Sandy SLOSH model Hurricane Inundation Zones data made available by the New York State Office of Emergency Management (NYSOEM) and "land use" data obtained from the New York City Department of City Plan-ning, we have plotted (Figures 3.4–3.6) urbanization patterns in these hurricane-vulnerable zones of Staten Island, NY. Hurricane storm-surge zones are based on NOAA SLOSH model projections of verti-cal surge heights associated with Saffir-Simpson-Scale category 1–4 storms (NYSOEM). Our GIS analysis shows the progressive urbaniza-

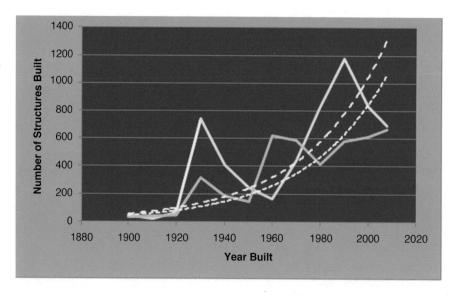

Fig. 3.4. Number of structures vs. year built on Staten Island for SLOSH Zones 1 and 2 from 1900 to 2008. White solid line = SLOSH Zone 1; yellow solid line = SLOSH zone 2; upper dashed line = trendline for SLOSH Zone 1; lower dotted line = trendline for SLOSH Zone 2. Data source: NYC PLUTO.

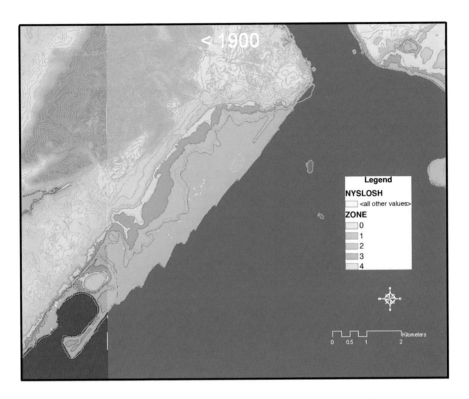

Fig. 3.5. GIS map showing east shore of Staten Island. The white squares indicate urbanized blocks and lots prior to 1900 appearing in SLOSH Zone 1.

tion in pre-Sandy SLOSH Zones 1 and 2. We have graphed these results in Figure 3.4. Note the exponential construction of structures despite the occurrence of hurricanes. Maps produced from this study, showing progressive urbanization, are very useful in analyzing the development of SLOSH zones on Staten Island. As a result of Superstorm Sandy, FEMA has revised the SLOSH zone maps and the New York City Office of Emergency Management revised its evacuation zones from three (A, B, and C) to six (1–6).

The damage plotted in Figures 3.7 and 3.8 is a result of the urbanization in the hurricane SLOSH zones as indicated in Figure 3.6. Note that most severe damage occurred in old tidal channels as shown in Figure 3.3.

Fig. 3.6. GIS map showing east shore of Staten Island. The white squares indicate urbanized parcels of land between the years 1898 and 2008 appearing in SLOSH Zones 1 and 2.

3.3 MODELING THE FUTURE OF SUPERSTORMS, BEFORE AND AFTER SANDY

For several years, we have been working as part of an interdisciplinary team using The City University of New York's Interdisciplinary High-Performance Computing Center (CUNY-IHPCC), housed at the College of Staten Island, to model how storm surges might impact the New York metropolitan area.

Our Center houses two of the latest-generation Cray supercomputers and covers all modern computational architecture. The CUNY-IHPCC can handle massive amounts of data.

Our computing power was also used by New York City for risk analysis and storm-surge prevention (for example, ARCADIS in Mayor

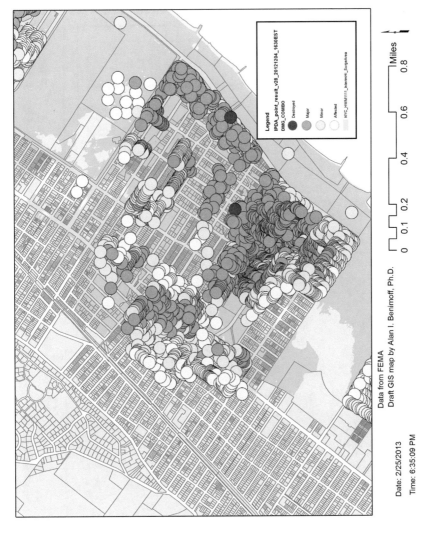

Fig. 3.7. FEMA damage assessment for Midland Beach Area. The blue area represents the Sandy inundation (source: FEMA). Note that the most severe damage occurs in old tidal channels shown in Fig. 3.3.

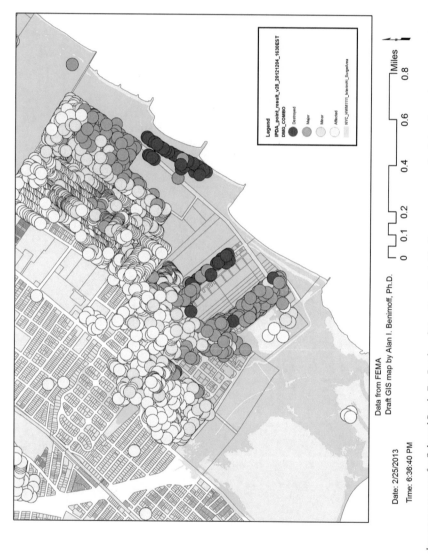

DAMAGE -Fox Beach - Oakwood Beach -

Legend
IPDA_point_result_v28_20121204_1630EST
DMG_COMBO
  Destroyed
  Major
  Minor
  Affected
  NYC_HWM111_Intensit_SurgeArea

Miles
0   0.1 0.2   0.4   0.6   0.8

Data from FEMA
Draft GIS map by Alan I. Benimoff, Ph.D.

Date: 2/25/2013
Time: 6:36:40 PM

Fig. 3.8. FEMA damage assessments for Oakwood Beach, Fox Beach, and New Dorp Beach. The blue area represents Sandy inundation (source: FEMA). As in Fig. 3.7, note that the most severe damage occurs in old tidal channels shown in Fig. 3.3.

Bloomberg's report on NYC special initiative for rebuilding and resilience [June 2013]).

3.3.1 Storm-Surge Applications for NYC

Many teams have modeled potential storm-surge impact from North Carolina south. However, few have looked at conditions north of the Carolinas. Because we have information from the sea floor north of the Carolinas, especially for New York Harbor, we formed a partnership with the Renaissance Computing Institute (RENCI) at the University of North Carolina and together have one of the most complete datasets of sea floor type, water depths, topography, and atmospheric conditions from the Gulf of Mexico to Canada. Using the CSI-IHPCC, we can model vulnerable areas along the entire Eastern Seaboard, again a massive dataset.

3.3.2 Response to Irene

After Hurricane Irene in August 2011, the CSI team was concerned that many people were lulled into a false sense of security. Because of the eye track, Irene was mostly a rain event with little wind or storm surge. Rain-induced flooding generally happens slowly, faster flooding is confined to stream bottoms where the Bluebelt could not handle the downpour. Concerned that people were not prepared for the onrush of a storm surge from the sea, we decided to model what a surge would look like with an eye track a little farther north than Irene and one that occurred on a high tide.

In June 2012, five months before Sandy, we wrote (Benimoff et al., 2012) that while most people do not think of New York as lying within the hurricane belt, powerful storms have impacted our city before, and we need to safeguard our communities. In addition to years of basic geologic fieldwork, we used data collected from Irene to model the impact of a storm that might hit at high tide with a slightly different track than Irene and calculated the likelihood of a 3.66-m surge. While we did not predict Sandy, we accurately modeled a scenario very close to that of Sandy.

3.3.3 Sandy Hindsight Simulation Using ADCIRC

Our initial model was based on data from Hurricane Irene using the Advanced Circulation Model (ADCIRC) numerical code and years of

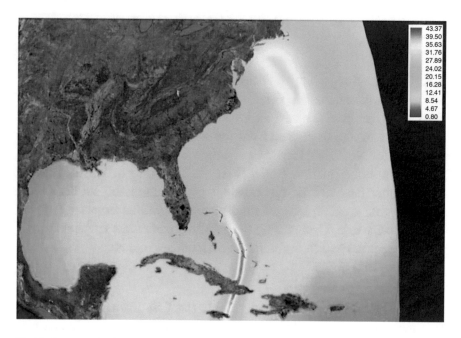

Fig. 3.9. Superstorm Sandy maximum wind speed in m/s along actual hurricane track as generated by the CUNY Interdisciplinary High-Performance Computing Center at the College of Staten Island.

fieldwork. However, following Superstorm Sandy, we have taken actual data from the storm and entered it into our model (Figure 3.9). As a result of our model, which used hindsight simulation, we arrived at surge simulation within 0.02 m of the actual observed event in many key locations. We are currently refining our numerical grid to better represent topographic and bathymetric detail to more accurately simulate storm surges to understand the potential impact of future storms on New York Harbor.

Using a GIS and data from FEMA, USGS, New York City, and New York State, we have also plotted (1) maximum surge, (2) building footprints, (3) damage from Sandy, (4) wetlands, (5) Bluebelt drainage basins, (6) urbanization patterns, (7) hurricane evacuation zones, (8) SLOSH zones, (9) elevation data, (10) FEMA flood zones, and (11) census data. The results of this study are very useful in learning from the impact of Sandy.

Using ADCIRC, in collaboration with RENCI, we modeled the storm surge. As shown in Figure 3.10, in many key locations on the

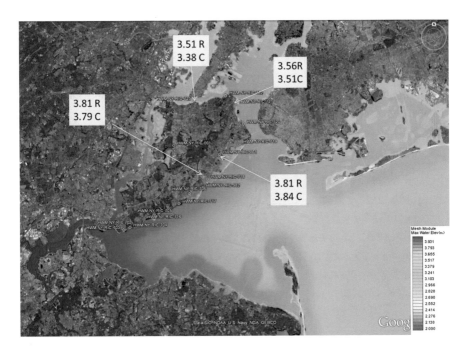

Fig. 3.10. Sandy surge showing good agreement in R – recorded and C – calculated values of surge. Units are meters. The Sandy inundation is shown in blue on Staten Island.

southeastern shore of Staten Island, agreement was within 0.02 m of the high-water mark recorded by the USGS. We now plan to use this refined model to understand the effectiveness of various flood-mitigation strategies, including artificial levees and bioengineered structures.

3.3.4 Potential for Real-time Prediction

Employing the numerical model used for Superstorm Sandy, along with expertise in forecasting from our RENCI partners, we are at the point of being able to make real-time predictions and models as an actual event develops. This prediction could give the New York area at least a few days' notice to prepare for an actual event and obtain block-level accuracy in storm-surge predictions. It is our hope that more accurate predictions will encourage people to act and move from harm's way.

In the future, we need to model how various engineered structures and bio-structures modify the surge, in regard to both mitigation and augmenting. Engineered structures often protect one area at the expense of another.

3.4 EVACUATION PROCEDURES

The maximum of the maximum envelope of high water (MEOW) or MOM provides a worst-case snapshot for a particular storm category under "perfect" storm conditions. The MOMs are also used to develop the nation's evacuation zones (source: FEMA).

3.5 RISK AND RESILIENCE

Coastal zone residents on Staten Island are at the highest risk of being affected by storm-surge inundation. The New York City Office of Emergency Management (NYCOEM) has redrawn its evacuation zone map (Figure 3.12). This map shows that it is possible to have an Evacuation Zone 1 adjacent to an Evacuation Zone 6 because of differences in topography. As a further result of Superstorm Sandy, FEMA has issued new SLOSH MOM maps (Figures 3.11 and 3.12).

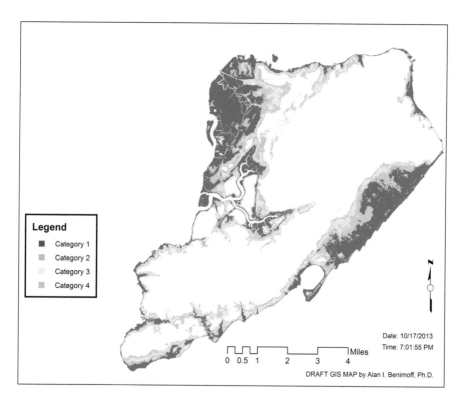

Fig. 3.11. 2012 FEMA SLOSH MOMS.

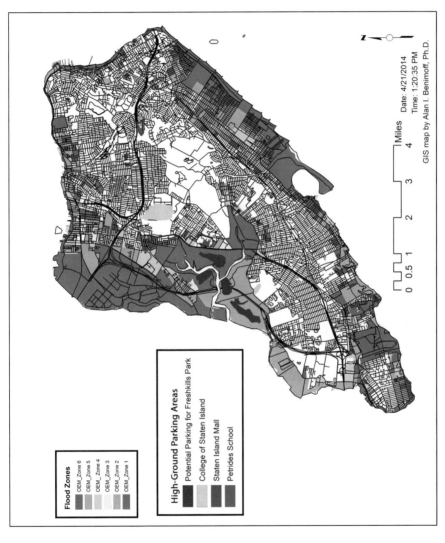

Flood Zones

- OEM_Zone 6
- OEM_Zone 5
- OEM_Zone 4
- OEM_Zone 3
- OEM_Zone 2
- OEM_Zone 1

High-Ground Parking Areas

- Potential Parking for Freshkills Park
- College of Staten Island
- Staten Island Mall
- Petrides School

Miles
0 0.5 1   2   3   4

Date: 4/21/2014
Time: 1:20:35 PM
GIS map by Alan I. Benimoff, Ph.D.

Fig. 3.12. Map showing high-ground parking areas and the six new coastal evacuation zones (NYCOEM).

Once evacuation orders are issued, it is important for coastal residents to seek high ground. We have delineated several high-ground parking areas that could be used by Staten Island coastal residents. Based on past hurricanes, the storm surge is usually gone within 0.5 day of its peak.

How one responds to a hurricane evacuation order is a matter of how that person perceives the order (Elsener and Kara, 1999). Each hurricane is different. For the Gulf Coast, Camille was different from Katrina. For New York City, Irene was different from Sandy. When the hurricane warning for Irene was issued, those who evacuated did so only to learn that the threat was minimal. Due to their past experience, many Staten Island residents did not perceive the seriousness of the evacuation order for Sandy.

In contrast, during Sandy 23 residents of Staten Island lost their lives, many lost their homes, and numerous motor vehicles were caught in the inundation as a result of Sandy's storm surge. The locations of twenty of the 23 who died are shown in Figure 3.13. Note the proximity of 11 fatalities to the stream in the New Creek Bluebelt (NYCDEP). Also note the paucity of contour lines. We hypothesize that the storm surge came up New Creek and inundated this area. The Bluebelt streams were there to drain storm water from this low-lying area.

3.6 A FIVE-POINT PLAN FOR THE FUTURE

In this chapter, we have attempted to show that the New York Metropolitan area is very vulnerable to hurricanes and tropical storms in part because of the unique geometry of the coastline. While tropical storms make it this far north less frequently than areas of the coastline further south, those that do can have a devastating impact on this highly populated area. Sandy was a relatively minor event with a 4.267-m storm surge. However, there is scientific speculation that the area could encounter surges of 9, and perhaps even 11 m if we were to be hit with a category 3 or higher storm with the right combination of tides, winds, and eye track.

Unfortunately, many coastal areas throughout New York Harbor have been "hardscaped"; old marshes, tidal channels, and dune fields have given way to urbanization. Because of the intense development in the low-lying areas of Staten Island and other areas throughout the city, protection and mitigation solutions are very different from coastal areas

Fig. 3.13. Portion of the narrows NY-NJ quadrangle showing fatalities (circles) from Sandy in this area. As in Figures 3.7 and 3.8 showing the most severe damage, the fatalities follow the old tidal channels shown in Figure 3.3. New Creek Bluebelt lies between 11 and the two clusters of circles of fatalities that lie along the coast immediately to the southeast.

that are largely recreational and consist primarily of second homes. By contrast, these areas on Staten Island house the only residences of low- to modest-income families and are filled with a socially vulnerable population.

Restoration of dune fields that restore natural protection still allows flooding of marshlands, an impossible solution in a heavily urbanized area. The intense building in a city also makes setbacks, rezoning, and relocation many times more difficult than in less populated parts of the coast. While as geologists we prefer natural solutions, we also understand the realities of planning in a major city and think that multiple solutions need to be employed to protect life and property.

In an effort to guide future planning in New York City as it prepares for future severe weather events, and as a model for consideration in other urban areas, we propose the following Five-Point Plan.

1. Protect our existing dunes, marshes, wetlands, and barriers whenever possible.
2. Rebuild and restore coastal dune fields and marshes.
3. Consider rezoning high-risk areas for day use and recreational purposes. Even within the flood zone, some areas are more vulnerable than others. For example, the old marsh channels shown in Figure 3.3 controlled the most severe destruction of homes.
4. Consider appropriate use of seawalls, floodgates, and other engineering solutions. Once again, it is critical that we understand that engineering solutions almost always protect one area at the expense of another.
5. Above all, we need to educate people that in storm surges when the water starts to rise, it is too late to escape. People need to climb to safety and never take shelter in a basement that could fill with water and trap victims in seconds, but head to high ground. Evacuation orders must be taken seriously in order to avoid loss of life, and government officials should be mindful of evacuation plans for people with disabilities, the ill, the elderly, and people in hospice and home care. Education can also guide appropriate building codes and construction styles when decisions are made to rebuild. Appropriate low-cost ADA-compliant signage should guide residents to high ground, and let them know the vulnerability of their location. Local officials should also designate areas on high ground where residents can move their vehicles to protect them from the storm (Benimoff et al., 2014).

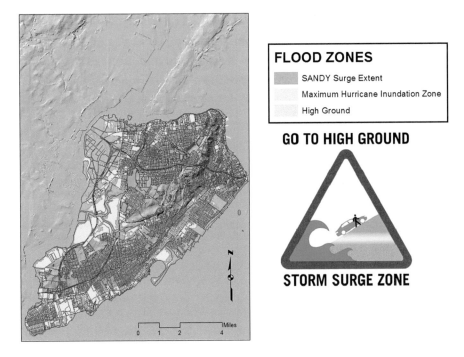

Fig. 3.14. "In case of hurricane storm surge go to high ground" Map. (Sources: NYC PLUTO, LION, USGS, FEMA, and NYCOEM).

Although our city is extremely vulnerable, we are much more fortunate than many other coastal cities because we have an ample amount of high safe ground (Figure 3.14). We need to take advantage of that fact, implement the five points listed above, and draw on the expertise and influence of a broad spectrum of scientific, political, social, and business experts to ensure the safety of everyone in New York City's hurricane-vulnerable zones.

ACKNOWLEDGMENTS

Helpful contributions were made by Brian O. Blanton of the Renaissance Computing Institute, University of North Carolina at Chapel Hill, Chapel Hill, NC, United States.

Contributions were also made by Paul Muzio, Eugene Dzedzits, and Liridon Sela of the CUNY Interdisciplinary High-Performance Computing Center, College of Staten Island, Staten Island, NY, United

States. The CUNY-IHPCC is operated by the College of Staten Island and funded, in part, by grants from The City University of New York, the State of New York, the CUNY Research Foundation, and National Science Foundation grants CNS-0958379 and CNS-0855217. We thank Terry Mares and Debbie Mahoney for assistance in preparation of the manuscript.

REFERENCES

Benimoff, Alan I, Fritz, William J., Kress, Michael, 2014. Staten Island severe storm evacuation: go to high ground and bring your automobile. Geological Society of America Abstracts with Programs 46 (2), 56.

Benimoff, A.I., Blanton, Brian O, Dzedzits, Eugene, Fritz, William J., Kress, Michael, Muzio, Paul, Sela, S Liridon, 2013. Storm Surge Modeling of Superstorm Sandy in the New York City Metropolitan Area. Fall 2013 Meeting. American Geophysical Union.

Benimoff, Alan I, Fritz, William J., Kress, Michael, 2013a. Geomorphic impact of Hurricane Sandy on Staten Island. Geological Society of America Abstracts with Programs 45 (1), 134.

Benimoff, Alan I, Fritz, William J., Kress, Michael, 2013b. Learning from the impact of Superstorm Sandy on Staten Island. Geological Society of America Abstracts with Programs 45 (7).

Benimoff, Alan I, Fritz, William J., Kress, Michael, 2013c. Learning from the impact of Superstorm Sandy on Staten Island NY. Geological Society of America Abstracts with Programs 45 (7), 51.

Benimoff, Alan I, Blanton, Brian O, Dzedzits, Eugene, Fritz, William, Kress, Michael, Muzio, Paul, 2012. Storm surge model for New York, Connecticut, and northern waters of New Jersey with special emphasis on New York Harbor. Geological Society of America Abstracts with Programs 44 (7).

Benimoff, Alan I, 2012. Hurricane history, extratropical storms, nor'easters and flooding for Staten Island, NY: a review, hydrologic response and further GIS studies. Geological Society of America Abstracts with Programs 44 (2), 61.

Benimoff, Alan I, 2010. A GIS study of urbanization in hurricane SLOSH Zones on Staten Island. Geological Society of America Abstracts with Programs 42 (1).

Blake, E.S., Kimberlain, T.B., Berg, R.J., Cangialosi, J.P., Beven, I.I., J.L., 2013. Tropical Cyclone Report Hurricane Sandy. Report AL182012, National Hurricane Center, p. 157.

Elsener, James, B., Kara, A.B., 1999. Hurricanes of the North Atlantic. Oxford University Press, p. 488.

Hall, Timothy M, Sobel Adam H, 2013. On the Impact Angle of Hurricane Sandy's New Jersey Landfall. Geophysical Research Letters 40, 2312–2315, doi;10.1002/grl.50395.

Kieffer, Susan W, 2013. The Dynamics of Disaster. W.W. Norton, p. 315.

Rahmstorf, S., 2012. Modeling sea level rise. Nature Education Knowledge 3 (10), 4.

Scileppi, E., Donnelly, J.P., 2007. Sedimentary evidence of hurricane strikes in Western Long Island New York. Geochemistry, Geophysics Geosystems 8, Q06011, doi:10.1029/2006GC001463.

United States Army Corp. of Engineers, 1964. New York to Fort Wadsworth to Arthur Kill Cooperative Beach Erosion and Interim Hurricane Study (Survey).

CHAPTER 4

High Frequency Trends in the Isotopic Composition of Superstorm Sandy

Stephen P. Good, Derek V. Mallia, Elizabeth H. Denis, Katherine H. Freeman, Xiahong Feng, Shuning Li, Nicolas Zegre, John C. Lin and Gabriel J. Bowen

ABSTRACT

Reliable forecasts of extra-tropical cyclones such as Superstorm Sandy require accurate understanding of their thermodynamic evolution. Within such systems, the evaporation, transport, and precipitation of moisture alters stable isotope ratios of cyclonic waters and creates spatio-temporal isotopic patterns indicative of synoptic-scale process-es. Here, high-frequency records of precipitation isotope ratios from four sites (West Lebanon, NH; Baltimore, MD; State College, PA; and Colcord, WV) are used to investigate the development of Sandy as the storm made landfall and moved inland. These high-frequency records are also combined with a Lagrangian backward transport model to create a general relationship between precipitation deuterium-excess and moisture source conditions. Based on this general relationship, the evolution of precipitation efficiency within Superstorm Sandy is mapped through time using a set of distributed isotope collections. These maps identify a region of high-precipitation efficiency near storm's core where intense rainfall rates likely exceeded the resupply of moisture as well as outlying rain-bands of lower precipitation efficiency possibly influenced by entrainment of a mid-western cold front.

Keywords: isotope; Lagrangian; deuterium-excess; Rayleigh distillation; precipitation efficiency

Learning from the Impacts of Superstorm Sandy. http://dx.doi.org/10.1016/B978-0-12-801520-9.00004-3

4.1 ISOTOPES AS TRACERS OF CYCLONIC PROCESSES

The stable isotope composition ($\delta^{18}O$ and δ^2H) of precipitation and water vapor collected during hurricanes has been shown to record information about the source, evaporation, transport, and condensation of moisture within large dynamic systems (Gedzelman and Lawrence, 1982, 1990; Lawrence and Gedzelman, 1996; Lawrence et al., 1998, 2002; Gedzelman et al., 2003; Coplen et al., 2008; Fudeyasu et al., 2008; Good et al., 2014). During Superstorm Sandy, a uniquely large dataset of cyclonic precipitation isotopic composition was assembled through a crowdsourcing approach and used to assess the spatio-temporal evolution of the storm. A broad overview of the precipitation collection locations, analysis methodology, and storm-wide spatio-temporal isotope patterns is presented in Good et al. (2014).

Of particular interest within this dataset are four sites at which high-frequency collections of precipitation were conducted. From east to west, these sites are: West Lebanon NH, Baltimore MD, State College PA, and Colcord WV. Samples from these sites geographically and temporally span the extent of the storm, and document isotopic features indicative of complex processes that occurred as Sandy evolved. In this chapter, we present an analysis of the isotopic trends at these sites. Additionally, a state of the art Lagrangian back-trajectory model was used to construct a general relationship between isotopic composition and the extent of precipitation rainout. This relationship is then applied to the dataset of 600+ samples from Good et al. (2014) to assess storm wide patterns in Sandy's thermodynamic evolution. The spatial and temporal distribution of precipitation collected at these four locations and all other sites is shown in Figure 4.1.

4.1.1 Background on Stable Isotopes in Precipitation

Stable isotopes have been used to investigate cyclones since the 1980s, with considerable early research conducted by Stanley Gedzelman and James Lawrence (Gedzelman and Lawrence, 1982, 1990). These studies investigated the stable isotope composition of oxygen, $\delta^{18}O$ [‰], and hydrogen, δ^2H [‰] ($\delta = R_{sample}/R_{std}-1$ where $R = {}^2H/{}^1H$ or ${}^{18}O/{}^{16}O$), and focused on understanding factors that influenced the isotopic composition of precipitation collected during hurricanes.

Broadly speaking, the stable isotope ratio of precipitation during cyclones is lower than typical mid-latitude precipitation (Gedzelman

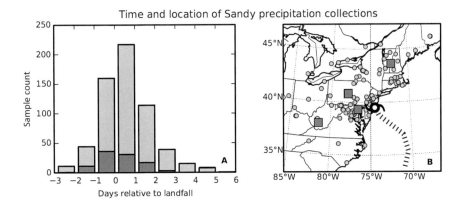

Fig. 4.1. (A) Histogram of the number of precipitation samples as a function of their collection time relative to landfall (yellow bars). (B) The location of all precipitation collected during Sandy (yellow circles). The four locations where high frequency precipitation samples were collected (green bars and squares) correspond to the sites where STILT back trajectories were initiated.

and Lawrence, 1982, 1990). Coherent spatial patterns have been found in cyclonic rainwater isotope values and vapor samples (Lawrence and Gedzelmanm, 1996; Lawrence et al., 1998, 2002), with isotope ratios near the eyewall being the most isotopically depleted. Stable isotope samples have also been used to assess the hydrologic budget of storms, the strength of circulation, and shifts in updraft velocities (Lawrence et al., 2002, Gedzelman et al., 2003, Coplen et al., 2008, Fudeyasu et al., 2008).

4.1.2 Rayleigh Fractionation

A first-order interpretation of precipitation isotopic composition can be derived from a consideration of a Rayleigh distillation process (Rayleigh, 1902). As an air mass progressively loses moisture, water enriched in ^2H and ^{18}O is more likely to condense, leaving the remaining air mass depleted in these isotopes (i.e. lower δ values). Isotopic fractionation during condensation ($\alpha = R_{rain}/R_{vapor}$) is assumed to be a temperature dependent equilibrium process (Dansgaard, 1964; Majoube, 1971). The isotopic depletion of cloud vapor from time t to $t + 1$ is expressed as

$$\delta_{t+1} = \delta_t + (\alpha - 1)\ln\left(q_{t+1} / q_t\right),\tag{4.1}$$

where q is the cloud vapor mixing ratio and $q_t - q_{t+1}$ is the amount of moisture that has condensed during a short time interval. We define $1 - q_t/q_0$

as the Rayleigh rainout fraction, where this quantity describes the fraction of original source water lost to condensation processes during the thermodynamic evolution of an air parcel. The Rayleigh approach assumes the instantaneous removal of condensed liquid with α calculated iteratively based on the dew point temperature of an adiabatic rising air mass (Dansgaard, 1964; Gedzelman et al., 2003).

The initial isotopic composition, δ_0, is dependent on the temperature and relative humidity at the source of evaporation, with δ_0 modeled using the closure assumption of Merlivat and Jouzel (1979) (evaporated water is isotopically equal to oceanic surface vapor). Thus, δ_0 is determined based on the sea surface temperature, T_0, and ambient humidity normalized to the saturation vapor pressure at the surface, h_0, at the source evaporation location as

$$\delta_0 = (1-k)/\left(\alpha(T_0)(1-k\,h_0)\right) - 1, \tag{4.2}$$

where k is the kinetic fractionation factor ($k_{\delta 18O} = 6.2\text{‰}$ and $k_{\delta 2H} = 5.5\text{‰}$) (Uemura et al., 2008). Because of differences in the molecular diffusivity among $H_2^{16}O$, $H_2^{18}O$ and HDO, a strong relationship exists between the relative humidity and deuterium excess ($d = \delta^2H - 8\,\delta^{18}O$) whereby evaporation under lower humidity results in vapor with higher d values (Merlivat and Jouzel, 1979).

Once δ_0 is estimated based on initial conditions (T_0 and h_0), Eqn. (4.1) is applied iteratively until a final cloud vapor composition equal to the value that would produce the observed rain δ value is reached. The ratio of the final cloud-mixing ratio to the initial cloud-mixing ratio (q/q_0) is then used to calculate the Rayleigh rainout fraction.

4.1.3 Precipitation Efficiency In Cyclones

The Rayleigh rainout fraction can be used to estimate how efficiently convection produces precipitation. This quantity is also known as precipitation efficiency as defined in Sui et al. (2007). Past studies have estimated precipitation efficiency in an effort to forecast and diagnose convective systems that produce heavy rain and flash flooding (Doswell et al., 1996).

Precipitation efficiency is also a key parameter for several cumulus parameterization schemes utilized in operational numerical weather prediction models today (Kuo, 1965; Kain, 1993; Grell, 1993). There are also cloud-climate feedbacks where precipitation efficiency is an important parameter, which can have serious implications for climate models (Lau and Wu, 2003). A better understanding of Rayleigh rainout or precipitation efficiency is needed in order to better resolve weather and climate-scale processes within numerical models.

4.2 HIGH FREQUENCY ISOTOPE RECORDS COLLECTED DURING SUPERSTORM SANDY

The sites of West Lebanon, New Hampshire (17 samples, −72.30E, 43.64N); Baltimore, Maryland (27 samples, −76.62E, 39.34N); State College, Pennsylvania (18 samples, −77.89E, 40.80N); and Colcord, West Virginia (37 samples, −81.43E, 37.94N) were selected as the focus of this investigation. At each of these sites, a large number of samples were collected at higher temporal frequencies, and collectively these sites are distributed widely around Sandy's core.

Precipitation collected at these (and other) locations was analyzed at the University of Utah, with the methodology and analytical precisions described in Good et al. (2014). Figure 4.2 depicts the isotopic time series of precipitation and the deuterium excess (d) at the four sites. Also shown with the isotopic information are the Stage IV radar hourly precipitation rates obtained from the NOAA National Center for Environmental Precipitation (NCEP) (Lin and Mitchell, 2005). These radar observations have a spatial resolution of 4 km, and thus provide precipitation estimates closest to collection locations. However, radar rainfall estimates have a minimum observable rainfall rate (~0.01mm/h), and during some periods rainfall was collected though not reported in the Stage IV dataset.

4.2.1 West Lebanon, New Hampshire

Samples from West Lebanon consisted of both 12-h and higher frequency collections. The 12-h samples were collected using a 10 cm funnel feeding into a 1 L Nalgene bottle through a drill hole on the cap. At the end of each 12-h collection period, the bottle was replaced with another dry bottle, and immediately capped with a regular airtight cap. The high frequency samples were collected using a plastic bucket with an opening of 16 cm in diameter. The sampling frequency was determined by the rainfall rate and logistical constraints, and at the end of each collection period, the bucket was replaced by another dry bucket. For each

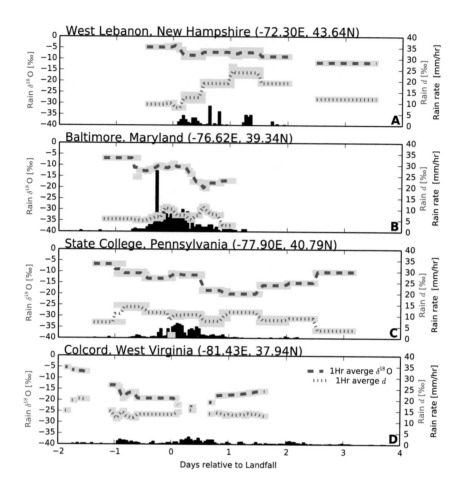

Fig. 4.2. High frequency time series of precipitation isotopic composition (dashed blue), deuterium excess (dotted red), and rainfall rate (black bars) at the four sites. Light colored blocks represent collections with specific start and end times and darker lines are 1-hour average estimates.

sample, 30 ml of water was transferred to a smaller bottle, which was refrigerated until its shipment.

West Lebanon is the most Eastern and Northern focus site in this study. As such, precipitation arrived at this site later than the other sites. The average δ^{18}O isotopic composition of precipitation was −6.9‰, with an average d of 15.0‰. The δ^{18}O value at this site varied the least among the four sites, with a range of only 7.6‰. Based on the radar precipitation data, rainfall arrived at this site in a series of distinct bands, and our data show that each band generally became isotopically lighter than the previous one. The d value at this site during the early portion of the storm was 9.2‰, and then increased significantly as the storm moved inland, resulting in a wide range of values (18.7‰). The maximum d value of 26.5‰ was the highest observed at the four sites. These constitute some of the highest d values found throughout the storm, with the peak occurring around 1 day after landfall as the storm began to break up. As with the $\delta^{18}O$ values, d values changed with each arriving rain-band.

4.2.2 Baltimore, Maryland

High frequency precipitation samples were collected as Sandy passed over the Baltimore area. Samples were collected hourly between 08:30 on October 29 and 14:00 on October 30, except for the periods from 20:34 on October 29 to 02:00 on October 30, and from 04:02 to 07:05 on October 30, when several hours passed between water collections. All samples from Baltimore were collected using a 2 L polypropylene bottle coupled with a plastic funnel. A funnel with a fine mesh cover was used to filter out leaves or other debris, and a ping-pong ball was situated in the funnel to prevent evaporation during rainwater accumulation. Water samples were stored in 1 dram glass vials, which were overfilled, capped with polycone seal caps, and sealed with Parafilm®, and then stored in a cool, dark environment until the time of analysis.

The Baltimore site is located closest to where Sandy came ashore, and experienced some of the highest rainfall rates during the storm. The average $\delta^{18}O$ isotopic composition of precipitation at this site was −13.7‰, with an average d of 7.5‰. Early precipitation was relatively ^{18}O-enriched at −6.6‰, and as the storm made landfall the $\delta^{18}O$ composition dropped to −13.5‰. The $\delta^{18}O$ value remained relatively steady during the heaviest rainfall, and then began to decrease once the rain rate slowed, dropping from −10 to −20‰. As the storm moved off to the west, the $\delta^{18}O$ values reached their minimum of −20.5‰ at approximately 12 h after landfall. The deuterium excess values at the Baltimore site were considerably lower than at the New Hampshire location. Additionally, the total variability was smaller, with a range of only 8.5‰. From approximately 12 h before storm landfall to approximately 12 h after landfall, the d and $\delta^{18}O$ values were positively correlated. After this point the d became negatively correlated with the $\delta^{18}O$, and the d reached its maximum of 11.4‰ when the $\delta^{18}O$ value reached its minimum.

4.2.3 State College, Pennsylvania

Samples from State College consisted of both 12-h samples and high frequency collections. Samples were obtained from plastic buckets placed outside in an open area. At collection time the contents of the bucket were poured into a collection container, either a 500 mL plastic bottle or 20 mL scintillation vial. Any remaining water in the bucket was dumped out before the bucket was set out for the next collection. The collection vials were labeled, sealed and refrigerated until they were shipped. Samples collected in larger plastic bottles were transferred to scintillation vials for shipping and sealed with Parafilm®.

The State College site is located approximately 200 km inland from the Baltimore site. As such, precipitation reached its peak intensity 9 h after the peak rainfall intensity was observed in Baltimore. Overall, the $\delta^{18}O$ value at this site showed a similar trend to that in Baltimore, with values starting at −6.6‰ before landfall and reaching their minima of −20.4‰ one day after landfall. In both the State College and Baltimore sites, a slight increase in $\delta^{18}O$ composition was observed during periods of higher rainfall around the landfall (approximately 12 h before to 12 h after). Similar to the Baltimore site, d values at State College ranged between 1.3‰ and 15.2‰ with an average of 9.7‰. A single sample collected during landfall had a very low d value (1.3‰). The $\delta^{18}O$ and d values were positively correlated through 12 h after landfall, and negatively correlated after this time.

4.2.4 Colcord, West Virginia

As part of a larger study evaluating streamflow generation processes in headwater catchments in the central Appalachian Mountains, automated water samplers (model 3700, Teledyne ISCO, Inc., USA) were used to collect precipitation at Colcord, WV. ISCO bottles were lined with 2 oz Nasco sample bags pre-treated with mineral oil to prevent fractionation from evaporation. Based on communication from individuals at the collector location on 29th, precipitation changed to snow sometime that evening, after which about 6″ of snow fell overnight and into the next day. From the afternoon of 30th onward, it changed to rain or a mix of rain and snow. The snow from the evening of the 29th through the 30th was likely collected in the funnel and was not released into the ISCO until it was melted by rising temperatures and/or rainfall. Samples taken on the afternoon of the 30th were probably an amalgamation of the snow that fell on the evening of the 29th and the morning of the 30th, and possibly the rain from that afternoon. Thus, samples collected from landfall (7:30 on the 29th) through approximately 16.5 h after landfall (mid-day on the 30th) should be regarded as possibly time-averaged, and viewed with caution.

Colcord is the Western-most location studied here in detail. As such, the precipitation recorded at this location before the landfall likely originated from a cold front moving eastward out of the mid-western United States. The earliest precipitation associated with this site was ^{18}O-enriched relative to the bulk of the Sandy-associated waters. However, after about 1 day before landfall, $\delta^{18}O$ values dropped considerably from $-5.4‰$ to $-20.0‰$. Post-landfall, the $\delta^{18}O$ value reached its minimum ($-23.1‰$), and then slowly began to rise. The trend in deuterium excess at the Colcord site is indicative of a mix of precipitation from the mid-latitude trough and Sandy associated waters. Early d values at this site were relatively high, with a maximum at 20.5‰. However, post landfall, the d values dropped to 11.4‰ and remained steady at these levels.

4.3 LAGRANGIAN BACK-TRAJECTORY ANALYSIS

In order to assess the isotopic evolution of cyclonic moisture associated with Superstorm Sandy, information is needed on the isotopic composition of source moisture (δ_0). Lagrangian back-trajectory analysis has been demonstrated to be a useful tool for constraining the origin of atmospheric water vapor (Soderberg et al., 2013; Brown et al., 2013; Good et al., 2014). Here we use the stochastic time-inverted Lagrangian transfer (STILT) model to estimate the source vapor isotopic composition for each hour that precipitation was collected at the four focus sites.

4.3.1 The Stochastic Time-Inverted Lagrangian Transfer (STILT) Model

The STILT model was used to generate backward trajectories in order to determine the origin of the water collected at the four focus sites. The STILT model is a state of the art backward Lagrangian particle dispersion model that uses an ensemble of passive particles to derive the upstream influence region for some specified receptor (Lin et al., 2003). Each particle within the STILT ensemble represents atmospheric transport that moves quasi-stochastically within the planetary boundary layer in response to turbulent eddies, which is treated as a Markov chain process (Lin et al., 2003; Wen et al., 2012; Kim et al., 2013; Lin, 2013). The particles within the STILT model were advected based on the gridded wind fields obtained from the NOAA-NCEP North American Mesoscale final analysis (NAM-FNL; Janjic, 2003). Temperature, pressure, and specific humidity from the NAM-FNL are interpolated to each particle location in time and space.

STILT model simulations were carried out at the four different receptor locations for sampling periods between October 27 and November 4. Five particles were released every 15 min at each of these locations in order to cover the entire collection period. Each particle simulation went backward 3 days with 15-min time steps. Sets of 5 particles were released from heights of $z = 500$ m, 3500 m, 6500 m, 9500 m, 12,500 m, and 15,500 m above the ground since the exact level of precipitate condensation height is unknown.

4.3.2 Using STILT to Estimate Source Vapor Conditions

The STILT model quantifies upstream influences for some receptor location x_r and time t_r using the unit flux footprint $f(x_r, t_r | x_i, t_i)$ which measures the surface influence (Lin et al., 2003; Wen et al., 2012; Kim et al., 2013; Lin, 2013). The footprint is a function of the number of Lagrangian particles within the planetary boundary layer at some upstream location and has units of mixing ratio per surface flux. The direct contribution of the upstream region can then be estimated by multiplying the footprint by the evapotranspiration moisture flux, and from these estimates, the flux-weighted initial vapor temperature (T_0) and relative humidity (h_0) were derived and used to model the source vapor isotopic composition (Good et al., 2014).

Using the STILT back trajectories we estimate T_0 and h_0 for each hour that precipitation was collected at the four locations. Figure 4.3(A) shows the calculated value of h_0 and the measured deuterium excess of Sandy precipitation based on precipitation collected from 18 h before landfall through 54 h after landfall (earlier times were excluded to remove the possible influence of the western cold front). The average value of T_0 was 17.2 ± 2.9°C, and the theoretical d calculated with Eqn. (4.2) using this temperature (dashed line, Figure 4.3(A)) matches closely the observed linear relationship between h_0 and collected d.

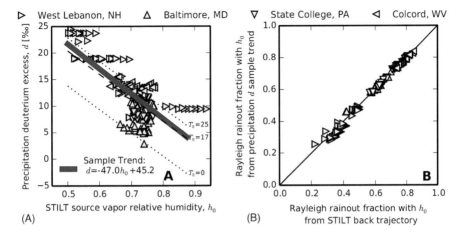

Fig. 4.3. (A) Deuterium excess at the four locations as a function of the flux-weighted surface normalized source vapor relative humidity (h_0) obtained from STILT back trajectory modeling. The sample trend (solid blue line) falls between d values expected from vapor with a source sea surface temperature (SST) of 0°C and 25°C (dotted lines) and near those expected based on the average source temperature, T_0, of 17°C (dashed line). Measured precipitation d values are used with the linear sample trend to approximate h_0, with T_0 assumed as 17°C. These source vapor conditions are then used with the measured $\delta^{18}O$ values to calculate the (B) Rayleigh rainout fraction and are compared with results obtained from STILT back trajectories.

4.3.3 General Relationship Between Source Conditions and Deuterium Excess

Based on the strength of the sample trend in Figure 4.3(A) and the low standard deviation of source surface temperature (2.9°C), we use the measured d values of collected precipitation to estimate the source evaporation humidity as

$$h_0 = (d - 45.2)/-47.0. \tag{4.3}$$

With Eqn. (4.3), the initial evaporation conditions are approximated based on d values and Eqn. (4.2) is then used to find the initial vapor isotopic composition. Because of possible errors associated with this assumption and the non-linearities involved in Eqn. (4.1), a Monte Carlo approach is used whereby 100 random source δ_0 values are estimated based on uncertainties in the sample trend and source temperature, and these δ_0 values are then used to estimate the Rayleigh rainout fraction for each precipitation sample 100 times. The average of these rainout fractions gives the expected rainout fraction using Eqns. (4.1)–(4.3). Figure 4.3(B) compares the Rayleigh rainout fraction calculated using the full STILT back trajectory initial conditions (T_0 and h_0) with Rayleigh rainout fraction calculated using initial conditions based on the Eqn. (4.3) and the average source temperature. The root mean squared error is 0.02 ($r^2 = 0.99$) between these two approaches.

The same approach of using the d value of collected precipitation to estimate source conditions is then applied to the entire dataset of Good et al. (2014). Figure 4.4(A–F) depicts the spatial pattern in rainout from 12 h before landfall to 48 h after landfall. Each panel includes all samples collected within 6 h of the specified time with the spatial rainout pattern estimated from the point values using a bilinear

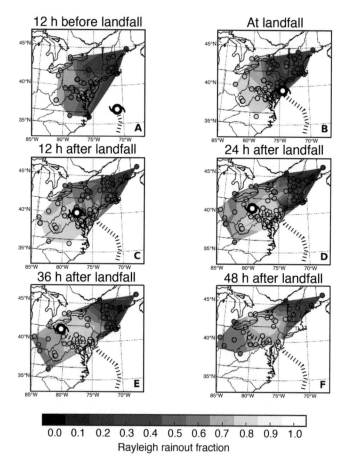

Fig. 4.4. *The spatio-temporal evolution of the Rayleigh rainout fraction during Superstorm Sandy based on data collected at all sites from Good et al. (2014).*

interpolation. Thus by using the deuterium excess in the collected precipitation to determine vapor source conditions, and by using the final isotopic composition of precipitation to estimate the strength of Rayleigh distillation, we can estimate the precipitation efficiency (or Rayleigh rainout fraction) for all precipitation collected during Superstorm Sandy.

4.4 EVOLUTION OF RAINOUT EFFICIENCY DURING SUPERSTORM SANDY

The reconstructed rainout fraction patterns provide insight into the interworking of Superstorm Sandy (Section 4.3.3 and Figure 4.4). Prior to and at landfall, the highest rainout fractions (>0.7) are located in the south and west of the study domain. These represent the downwind regions of the storm circulation, where air masses had longer over-land fetch. These isotope data suggest that rainout rates far exceeded the re-supply of moisture from the land surface along these circulation trajectories, and by the time air parcels have reached the southwest quadrant of the storm much of their vapor had precipitated out.

The highest rainout fractions were observed shortly after landfall, consistent with the peak in rain rates at coastal mid-Atlantic sites. At this time the spatial distribution of high-rainout sites expanded north and east of the storm center. As the storm moved inland, a broader area around the storm center was affected by circulation that had limited interaction with the Atlantic Ocean, and our data suggest this was reflected by expansion of high rainout fractions across a large part of the storm-affected area. Low fractions were sustained across much of New England, which primarily received rainfall from rain bands at the storm's periphery carrying high-d moisture attributed to continued evaporation off of the Atlantic Ocean (Good et al., 2014). From 24 h after landfall onward, as the storm began to weaken, rainout efficiency to the southwest of the core began to decrease. This suggests the possibility of resupply of moisture at the periphery of the storm, perhaps through entrainment of air from the west or through evaporation off of the Great Lakes.

In this study, the combination of both high-frequency and distributed isotope collections with STILT back trajectories presents a unique view into the water and energy cycling within Superstorm Sandy. The spatially resolved isotopic information shows strong trends throughout the evolution of the landfalling storm, contributing to our understanding of cyclone evolution, hydroclimatological impacts, and paleo-storm proxies.

ACKNOWLEDGMENTS

We gratefully thank all those who contributed samples to the collection effort. EHD and KHF acknowledge L.M. Patzkowsky, R.L. Oakes, and C.R. Magill for sampling assistance in State College, PA. Dan Davis, Crystal Tulley-Cordova, and Yusuf Jameel assisted with sample analysis. Support and resources from the Center for High Performance Computing at the University of Utah is gratefully acknowledged. This work was supported in part by US National Science Foundation Grant EF-01241286.

REFERENCES

Brown, D., Worden, J., Noone, D., 2013. Characteristics of tropical and subtropical atmospheric moistening derived from lagrangian mass balance constrained by measurements of HDO and H2O. Journal of Geophysical Research Atmospheres 118, 54–72.

Coplen, T.B., Neiman, P.J., White, A.B., Landwehr, J.M., Ralph, F.M., Dettinger, M.D., 2008. Extreme changes in stable hydrogen isotopes and precipitation characteristics in a landfalling pacific storm. Geophysical Research Letters 35, L21808.

Dansgaard, W., 1964. Stable isotopes in precipitation. Tellus 16 (4), 436–468.

Doswell III, C.A., Brooks, H.E., Maddox, R.A., 1996. Flash flood forecasting: An ingredients-based methodology. Weather and Forecasting 11 (4), 560–581.

Fudeyasu, H., Ichiyanagi, K., Sugimoto, A., Yoshimura, K., Ueta, A., Yamanaka, M., Ozawa, K., 2008. Isotope ratios of precipitation and water vapor observed in typhoon shanshan. Journal of Geophysical Research 113, D12113.

Gedzelman, S., Lawrence, J., Gamache, J., Black, M., Hindman, E., Black, R., Dunion, J., Willoughby, H., Zhang, X., 2003. Probing hurricanes with stable isotopes of rain and water vapor. Monthly Weather Review 131 (6), 1112–1127.

Gedzelman, S.D., Lawrence, J.R., 1982. The isotopic composition of cyclonic precipitation. Journal of Applied Meteorology 21, 1385–1404.

Gedzelman, S.D., Lawrence, J.R., 1990. The isotopic composition of precipitation from two extra-tropical cyclones. Monthly Weather Review 118 (2), 495–509.

Good, S.P., Mallia, D.V., Lin, J.C., Bowen, G.J., 2014. Stable isotope analysis of precipitation samples obtained via crowdsourcing reveals the spatiotemporal evolution of Superstorm sandy. PLoS One 9 (3), 1–7.

Grell, G.A., 1993. Prognostic evaluation of assumptions used by cumulus parameterizations. Monthly Weather Review 121 (3), 764–787.

Janjic, Z., 2003. A nonhydrostatic model based on a new approach. Meteorology and Atmospheric Physics 82 (1–4), 271–285.

Kain, J.S., 1993. Convective parameterization for mesoscale models: the Kain–Fritsch scheme. The representation of cumulus convection in numerical models. Meteorological Monograph 46, 165–170.

Kim, S.Y., Millet, D.B., Hu, L., Mohr, M.J., Griffis, T.J., Wen, D., Lin, J.C., Miller, S.M., Longo, M., 2013. Constraints on carbon monoxide emissions based on tall tower measurements in the US upper midwest. Environmental Science and Technology 47 (15), 8316–8324.

Kuo, H.-L., 1965. On formation and intensification of tropical cyclones through latent heat release by cumulus convection. Journal of the Atmospheric Sciences 22 (1), 40–63.

Lau, K., Wu, H., 2003. Warm rain processes over tropical oceans and climate implications. Geophysical Research Letters 30 (24.), .

Lawrence, J.R., Gedzelman, S.D., Gamache, J., Black, M., 2002. Stable isotope ratios: Hurricane Olivia. Journal of Atmospheric Chemistry 41 (1), 67–82.

Lawrence, J.R., Gedzelman, S.D., Zhang, X., Arnold, R., 1998. Stable isotope ratios of rain and vapor in 1995 hurricanes. Journal of Geophysical Research Atmospheres 103 (D10), 11381–11400.

Lawrence, R.J., Gedzelman, D.S., 1996. Low stable isotope ratios of tropical cyclone rains. Geophysical Research Letters 23 (5), 527–530.

Lin, J., 2013. Lagrangian Modeling of the Atmosphere, chapter Lagrangian Modeling of the Atmosphere, An Introduction. American Geophysical Union, AGU Geophysical Monograph.

Lin, J., Gerbig, C., Wofsy, S., Andrews, A., Daube, B., Davis, K., Grainger, C., 2003. A near-field tool for simulating the upstream influence of atmospheric observations: the Stochastic Time-Inverted Lagrangian Transport, (STILT) model. Journal of Geophysical Research Atmospheres 108 (D16), 4493–4506.

Lin, Y., Mitchell, K.E., 2005. The NCEP stage II/IV hourly precipitation analyses: Development and applications. In: Ninettenth Conference on Hydrology, number Paper 1. 2. American Meterological Socieity, San Diego, CA.

Majoube, M, 1971. Fractionnement en oxygene 18 et en deuterium entre leau et sa vapeur. Journal of Chemical Physics 68, 1423–1436.

Merlivat, L., Jouzel, J., 1979. Global climatic interpretation of the deuterium-oxygen 18 relationship for precipitation. Journal of Geophysical Research Oceans 84 (C8), 5029–5033.

Rayleigh, L., 1902. Lix. on the distillation of binary mixtures. The London, Edinburgh, and Dublin Philosophical Magazine and Journal of Science 4 (23), 521–537.

Soderberg, K., Good, S.P., O'Conner, M., Wang, L., Ryan, K., Caylor, K.K., 2013. Using atmospheric trajectories to model the isotopic composition of rainfall in central Kenya. Ecosphere 4 (3), 1–18.

Sui, C.-H., Li, X., Yang, M.-J., 2007. On the definition of precipitation efficiency. Journal of the Atmospheric Sciences 64 (12), .

Uemura, R., Matsui, Y., Yoshimura, K., Motoyama, H., Yoshida, N., 2008. Evidence of deuterium excess in water vapor as an indicator of ocean surface conditions. Journal of Geophysical Research Atmospheres 113, D19114.

Wen, D., Lin, J.C., Millet, D.B., Stein, A.F., Draxler, R.R., 2012. A backward-time stochastic lagrangian air quality model. Atmospheric Environment 54, 373–386.

CHAPTER 5

Hurricane Sandy: Did Beach Nourishment Save New Jersey?

Adam D. Griffith, Andrew S. Coburn, Katie M. Peek and Robert S. Young

ABSTRACT

The Program for the Study of Developed Shorelines (PSDS) at Western Carolina University (WCU), in conjunction with The Nature Conservancy, set out to test the hypothesis that New Jersey beaches nourished since 2000 provided a significant degree of storm damage reduction during Sandy. We identified and mapped all nourishment episodes in NJ between 2000 and 2012 and used mean high water (MHW) shoreline data from the NJ Bureau of GIS, county parcel data from the NJ Geographic Information Network, and version 28 of the publicly available FEMA Modeling Task Force (MOTF) data for structure damage estimates. Results of a logistic regression show a significant reduction in damage to structures behind beaches nourished since 2000, structures behind wider beaches, and structures located further from MHW. Additional research, such as an ordinary least squares regression, is recommended to determine the degree to which nourishment alone attenuated damage, which is currently unknown.

Keywords: Sandy; hurricane; storm damage; beach nourishment; developed shorelines; logistic regression; FEMA-MOTF

5.1 INTRODUCTION

When Hurricane Sandy slammed into New Jersey on October 29, 2012, it produced a major-to-record storm surge along much of the New Jersey coast due, in part, to the fact that landfall occurred near the time of astronomical high tide. Within days, anecdotal reports of reduced property damage behind nourished beaches garnered extensive media attention. Even though these reports were based on observation data from

Learning from the Impacts of Superstorm Sandy. http://dx.doi.org/10.1016/B978-0-12-801520-9.00005-5

Governor Chris Christie and other local elected officials and unsupported by objective data analyses, beach nourishment was quickly lauded by beach stabilization advocates as a panacea for developed shorelines in New Jersey (Spoto, 2012). Nourishment as a sustainable, long-term solution becomes problematic when large, slow-moving storms like Sandy directly impact heavily developed shorelines such as NJ, which eroded an average of 37 ft (Richard Stockton College, 2012a–f). Adopting a *de facto* policy of nourishment as a solution to accelerated erosion rates, especially in light of sea level rise, seems untenable at the national or state level.

United States beaches attract as many as 180 million visitors per year (Schwartz, 2005). The scale of the coastal travel and tourism industry attracts the attention of both local and national policy makers. The logic of a long-term, nourishment-based coastal strategy is questionable due to the cumulative costs of attempting to hold the nation's shorelines in place. This is of both local and national interest due to the large number of beach nourishment episodes that occur on the coast of the United States annually, and the public funds used to pay for them. Although nourishment is the favored method of attempting to keep erosion at bay, it is becoming more expensive as near-shore marine sand sources (borrow areas) are depleted.

After completing an aerial damage assessment of NJ and NY shorelines days after Sandy, PSDS, in conjunction with The Nature Conservancy of NY and NJ, set out to test the hypothesis that NJ beach nourishment projects completed since 2000 significantly reduced storm damage to beach front homes and buildings during Sandy, when compared to non-nourished beaches.

5.2 METHODS

Using the comprehensive PSDS US beach nourishment database, we identified and mapped all nourishment episodes in NJ between 2000 and 2012. We used mean high water (MHW) shoreline data from the NJ Bureau of GIS and county parcel data from the NJ Geographic Information Network. Focusing only on ocean beaches between Cape May and Sandy Hook in NJ, we selected the seaward-most structure (referred to as the front row) and mined data relating to the beach in front of that structure as related dependent variables. One initial problem we encountered was how to accurately assess Sandy-related damage due to the size of the study area and extensive number of impacted structures.

Although PSDS completed a reconnaissance-level aerial assessment of the NJ shoreline two weeks after Sandy, the only comprehensive, publicly available analysis of Sandy-related damage was FEMA's Modeling Task Force (MOTF) Hurricane Sandy Imagery Based Preliminary Damage Assessments (v28) compiled from NOAA aerial imagery, Civil Air Patrol photographs and media images. After major disasters, FEMA's Modeling Task Force (MOTF) uses remote sensing analysts to examine detailed satellite and aerial imagery to obtain estimates of damage, often for recovery planning and coordination purposes. In this case, FEMA-MOTF examined over 157,000 images taken between October 29 and November 18, 2012 (Federal Emergency Management Agency Modeling Task Force, 2013). The FEMA damage database includes point data representing individual assessments at the building level as well as damage classes for larger structures derived from a combination of imagery based assessments and modeled flood depth. Damage is broken into 4 categories: destroyed, major, minor, and affected using information on water depth, debris estimates, and site visits (Federal Emergency Management Agency Modeling Task Force, 2014). Each damage category is described as:

1. *Affected:* generally superficial damage to solid structures (loss of tiles or roof shingles); some mobile homes and light structures damaged or displaced
2. *Minor damage:* solid structures sustained exterior damage (for example missing roofs or roof segments); some mobile homes and light structures were destroyed, many were damaged or displaced.
3. *Major damage:* (wind) some solid structures were destroyed; most sustained exterior and interior damage (roofs missing, interior walls exposed); most mobile homes and light structures were destroyed. (Storm Surge) Extensive structural damage and/or partial collapse due to surge effects. Partial collapse of exterior bearing walls.
4. *Destroyed:* (wind) most solid and all light or mobile home structures destroyed. (Storm Surge) Structure was completely destroyed or washed away by surge effects.

Since the FEMA damage data are based on aerial imagery and inundation modeling, the actual damage inflicted by Sandy may be different than the FEMA damage database indicates. We believe, however, that any discrepancies between actual damage and the FEMA damage assessment are minor and do not impact the results of the analysis particularly in light of two facts: (1) our analysis includes the entire population of front row buildings, not a sample, and (2) that the same FEMA-MOTF team executed the damage estimates for the entire NJ coast (H.E. Longenecker, III, October, 23, 2013, personal communication). Table 5.1 below shows the sources of data used in this analysis.

Environmental Systems Research Institute (2012) ArcGIS 10.1 software was used to map the locations of damaged structures and corresponding features of interest including beach nourishment, ground elevation, dry beach width, distance to MHW and maximum dune height landward of MHW. Only first row structures were used due to their storm surge vulnerability and likelihood of benefit from beach nourishment. Figure 5.1 shows the distribution of FEMA-MOTF damage data. We also eliminated structures at inlets not directly facing the Atlantic Ocean due to highly complex inlet geomorphology.

Each structure centroid was assigned a representative point and was connected to the closest point on the NJ MHW vector line layer using a shortest distance polyline. This polyline also served as a profiler to determine dune elevation and distance to pre-Sandy MHW for each structure. To extract the maximum dune height corresponding to each structure, 200 random points were created along each profile line and elevation values were extracted to each random point using US Army Corps of Engineers pre-Sandy LiDAR data (2010), as shown in Figure 5.2. Distance from MHW was calculated as a simple attribute for each polyline.

Data from the NJ Beach Profile Network (NJBPN) of Richard Stockton College (2012a–f) were used to estimate beach widths. Their beach profile network provides detailed data on over 100 sites at roughly 1-mile

Table 5.1 Data and Sources Used in This Analysis	
Data	Source
Coastline of New Jersey	New Jersey Department of GIS (2012)
Damage Estimates	FEMA-MOTF v 28
Inundation	FEMA/USGS
NJ Beach Width	Richard Stockton College
NJ Elevation	ACE 2010 LiDAR
Beach Nourishment	PSDS

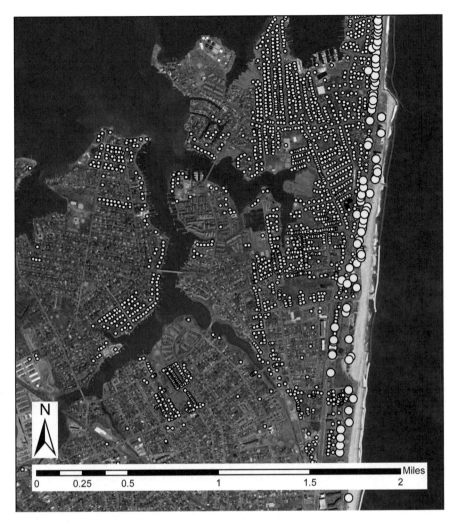

Fig. 5.1. FEMA-MOTF damage data points are shown in blue for Monmouth Beach, NJ. First row structures used in this analysis are highlighted in yellow. Green sections of the shoreline (MHW) indicate beaches not nourished since 2000 and red sections indicate beaches nourished since 2000. This section of beach is typical for the study area with varying beach widths between each structure and the shoreline.

intervals along the entire state. Beach width was determined by measuring the distance between the intersection of the beach profile and 0 feet NAVD and the point of the abrupt elevation change due to a boardwalk, fence, or dune on the heavily developed shoreline (Figure 5.3).

Ground elevation was extracted for all structure centroids using LiDAR data from the US Army Corps of Engineers LiDAR (2010)

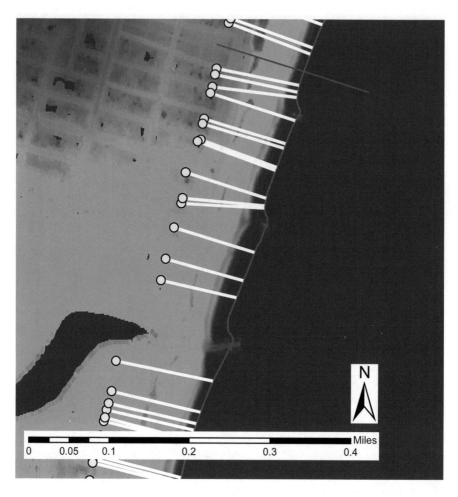

Fig. 5.2. *US Army Corps of Engineers LiDAR (2010) data is shown for the town of Ocean Grove, just south of Asbury Park. Damaged front row homes in yellow are connected to the coast (HWM) by white profile lines used to extract maximum dune heights. Elevations range from 22 ft in the N to 8 ft near Fletcher Lake in the SW. The blue polygons in the NW corner (and in other areas) are areas under buildings missing LiDAR data and are not lakes.*

accessed using NOAA Coastal Service Center's Digital Coast web portal. Structure centroids not located within an area of recorded LiDAR were manually relocated to the closest area of recorded LiDAR only for the purpose of extracting elevation values for these structures. Maximum dune elevation, distance to mean high water and elevation data were all imported into ArcGIS 10.1 with geographic coordinates associated with structure location. The PSDS US beach nourishment database was queried to identify and map nourishment episodes in

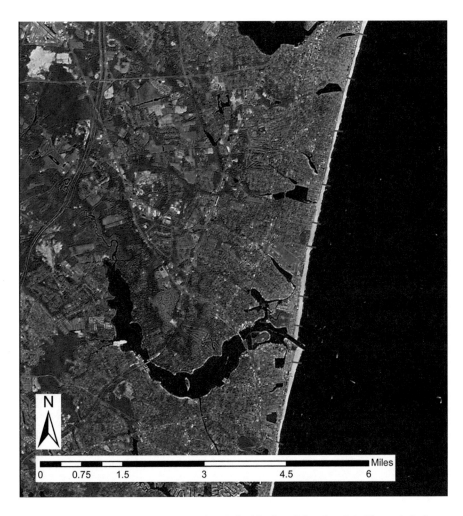

Fig. 5.3. Coast of NJ showing NJBPN transects from Richard Stockton College from Point Pleasant in the S to Asbury Park in the N. These transects were used to calculate beach width and damaged structures were assigned the beach width according to proximity.

NJ between 2000 and 2012. It is currently unknown how long nourished sand remains in a littoral system and 12 years was selected as a conservative estimate of this value. Focusing on New Jersey beaches between Cape May in the south and Sandy Hook in the north, our analysis shows that 58 beach nourishment episodes, covering approximately 45.5 miles (36%) of the 121-mile long developed New Jersey shoreline took place between 2000 and 2012 (Figure 5.4). Multiple

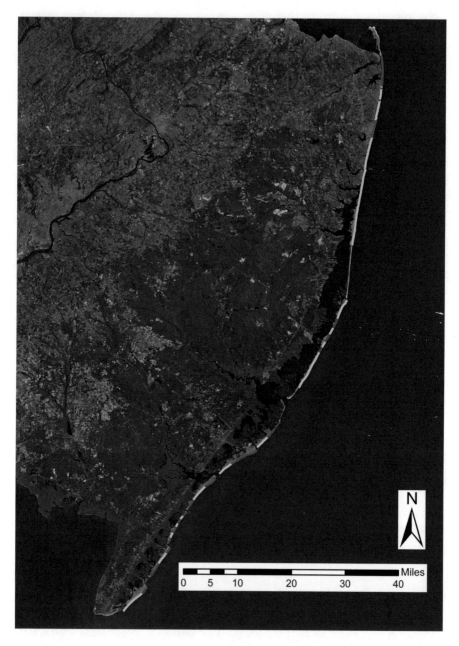

Fig. 5.4. The study area of this analysis is shown here: the Atlantic coast of New Jersey between Sandy Hook and Cape May shows beaches nourished since 2012 in red and non-nourished beach and beaches nourished prior to 2012 in green. The sum of the beaches studied is 112 miles, which excludes areas with no development and inlets.

beach nourishment episodes along the same stretch of beach were collapsed into a single category.

Statistical analyses were carried out in R 3.0.3 using a logistic regression generalized linear model for NJ. The null hypothesis is that there is no relationship between beach nourishment and reduced damage. Independent variables used in the statistical analysis were (1) the ground elevation of each structure, (2) the distance of each structure from MHW, (3) whether or not a structure was located behind a beach that had been nourished in the twelve years preceding Sandy and (4) maximum elevation of the ground between each structure and MHW. The field "DMG_VALUE" from the FEMA-MOTF version 28 expressed as a binary value (0 = no damage, 1 = damage) was used as the dependent variable.

5.3 RESULTS AND RECOMMENDATIONS

A total of 3788 first row structures in NJ were classified by FEMA as sustaining damage during Sandy. In terms of the degree of damage sustained, 3282 (86.64%) were classified as affected, 261 (6.89%) were classified as minor, 115 (3.04%) were classified as major and 130 (3.43%) were classified as destroyed. Of the 3788 damaged structures in the first row, 1345 (35.51%) were located behind beaches that had received beach nourishment between 2000 and 2012, while 2443 (64.49%) were located behind beaches that had not received beach nourishment during the same period.

Table 5.2 Damaged Structures in the First Row Relative to Beach Nourishment using FEMA-MOTF Damage Data			
Damage Level	No. of Structures	Behind Non-nourished Beaches	Behind Nourished Beaches
Affected	3282	2125	1157
Minor	261	144	117
Major	115	62	53
Destroyed	130	112	18
	3788	2443 (64.49%)	1345 (35.51%)

The total number of structures (Houses, Businesses, Detached Garages, etc.) in each FEMA-MOTF damage category is shown relative to its location behind a nourished or non-nourished beach. Note that the total number of damaged structures is lower in each damage category behind nourished beaches.

Upon initial logistic regression results, it became apparent that structural damage classified by FEMA as affected and minor was caused primarily by wind. Since beach nourishment does not mitigate wind damage, a decision was made to eliminate structures classified by FEMA as affected or sustaining minor damage from further analysis. After removing affected structures and structures receiving minor damage, 245 first row structures sustained major damage or were destroyed. Of these, 174 (71%) were located behind beaches that had not been nourished prior to Sandy, while 71 (29%) were behind beaches that had received nourishment.

When wind damage is removed from consideration, we see that the length of NJ shoreline considered nourished prior to Sandy (36%) is greater than the percent of first row structures located behind nourished beaches that were either destroyed or sustained major damage (29%) during Sandy. Similarly, 64% of non-nourished NJ shoreline accounted for 71% of first row structural damage. The statistical analyses show that the 7% differences in the descriptive statistics were significant.

Results of the logistic regression of the NJ shoreline indicate that first row structures located behind nourished beaches were less likely to be damaged than first row structures located behind non-nourished beaches. We rejected the null hypothesis that levels of damage were the same behind nourished and non-nourished beaches. An increase in each independent variable was shown to significantly reduce the likelihood of inundation: structures with higher ground elevations, and those farther from MHW, were less likely to be damaged.

Limitations on these results include several problems that lend themselves to recommendations for both future research and policy

Table 5.3 Summary of Logistic Regression Results for NJ Sandy Damage and Beach Nourishment

Coefficients:	Estimate	Std.	Error	Value	Pr (>|z|)
(Intercept)	4.077	0.179	22.749	2.00E-16	***
Beach Width	−0.003	0	−8.448	2.00E-16	***
Ground Elevation of Structure	−0.02	0.01	−2.124	0.0336	*
Distance to MHW	−0.003	0	−21.799	2.00E-16	***
Beach Nourishment	−0.436	0.081	−5.367	8.02E-08	***
Note: Significance Levels (*$p < 0.05$, **$p < 0.01$, ***$p < 0.001$)					

implications. Beach widths were calculated at approximately one-mile intervals based on the NJBPN transects and then assigned to the closest structures in that zone. True independence of observations of each structure was not achieved, and this possibly inflates the significance of beach width. Valuable future analyses would likely include an ordinary least squares (OLS) regression including a large variety of non-natural independent variables. Population demographics like annual household income often show significant relationships and an OLS regression would show the strengths of the relationships relative to one another. Although our regression results show a relationship between beach nourishment and reduced damage in New Jersey from Sandy, there are many other natural processes and anthropomorphic factors that contribute to or mitigate property damage during a large storm: along-shore variability in geology and geomorphology; along-shore variability in storm meteorology and oceanography; shoreline stabilization efforts and variability in the density, type, strength and elevation of buildings. Inclusion of these additional natural variables in an OLS regression would lead to a more robust analysis.

As a matter of national policy, however, beach nourishment as a viable solution to sea level rise seems impractical because of the high costs and inequitable distribution of costs and benefits. Program for the Study of Developed Shorelines (2014) reports total cumulative nourishment expenditures in excess of $2 billion from only four states: Florida, Georgia, North Carolina, and South Carolina. Before Sandy, NJ nourishment costs alone totaled $1 billion, making informed taxpayers wonder about the logic. The costs of nourishment are also increasing from $1.71 per cubic yard in the 1970s to over $14 per cubic yard now (Seabrook, 2013). Nourishment did reduce damage on the coast as a whole in NJ during Sandy, but in Sea Bright, not even $50 million of beach nourishment (Program for the Study of Developed Shorelines, 2014) could prevent 18 structures from being completely destroyed. Little could.

REFERENCES

Environmental Systems Research Institute, 2012. ArcGIS Software, Version 10.1. Redlands, CA.

Federal Emergency Management Agency Modeling Task Force 2013, FEMA Modeling Task Force (MOTF) Hurricane Sandy Response. Available from: <http://hazuscanada.ca/sites/all/files/hurricane_sandy_motf.pdf> (accessed 13.10.2013).

Federal Emergency Management Agency Modeling Task Force, 2014. FEMA MOTF-Hurricane Sandy Impact Analysis. Available from: <http://fema.maps.arcgis.com/home/item. html?id = 307dd522499d4a44a33d7296a5da5ea0> (accessed 11.04.2014).

McCallum, B.E., Wicklein, S.M., Reiser, R.G., Busciolano, Ronald, Morrison, Jonathan, Verdi, R.J., Painter, J.A., Frantz, E.R., Gotvald, A.J., 2013. Monitoring storm tide and flooding from Hurricane Sandy along the Atlantic coast of the United States, October 2012. U. S. Geological Survey Open-File Report 2013-1043.

New Jersey Department of Environmental Protection, GIS, 2012. Available from: <http://www. state.nj.us/dep/gis/stateshp.html> (accessed 27.03.2013).

Program for the Study of Developed Shorelines, 2014. Beach Nourishment Viewer. Available from: http://psds.wcu.edu/

Richard Stockton College of New Jersey, 2012a. Beach-Dune Performance Assessment of New Jersey Beach Profile Network (NJBPN) Sites at Atlantic County, New Jersey After Hurricane Sandy Related to FEMA Disaster DR-NJ 4086. Available from: <http://intraweb.stockton.edu/ eyos/coastal/content/docs/sandy/Atlantic.pdf> (accessed 13.02.2013).

Richard Stockton College, 2012b. An Assessment of Cape May County Beaches at the New Jersey Beach Profile Network (NJBPN) Sites After Hurricane Sandy Related to (DR-NJ 4086). Available from: <http://intraweb.stockton.edu/eyos/coastal/content/docs/sandy/Cape_May2.pdf> (accessed 13.02.2013).

Richard Stockton College, 2012c. Beach-Dune Performance Assessment of New Jersey Beach Profile Network (NJBPN) Sites at Long Beach Island, New Jersey After Hurricane Sandy. Available from: <http://intraweb.stockton.edu/eyos/coastal/content/docs/sandy/lbi_report.pdf> (accessed 13.02.2013).

Richard Stockton College, 2012d. Related to FEMA DR NJ 4086 Declared for Hurricane Sandy Beach-Dune Performance Assessment of New Jersey Beach Profile Network (NJBPN) Sites Between Deal and Sea Bright, New Jersey Related to FEMA DR-NJ 4086 Declared for Hurricane Sandy. Available from: <http://intraweb.stockton.edu/eyos/coastal/content/docs/sandy/northern-Monmouth.pdf> (accessed 13.02.2013)

Richard Stockton College, 2012e. Beach-Dune Performance Assessment of New Jersey Beach Profile Network (NJBPN) Sites at Northern Ocean County, New Jersey After Hurricane Sandy Related to FEMA Disaster DR-NJ 4086. Available from: <http://intraweb.stockton.edu/eyos/coastal/content/docs/sandy/NorthernOcean.pdf> (accessed 13.02.2013).

Richard Stockton College, 2012f. Beach-Dune Performance Assessment of New Jersey Beach Profile Network (NJBPN) Sites at Between Manasquan Inlet and Allenhurst, New Jersey Related to FEMA DR-JI 4086 Declared for Hurricane Sandy. Available from: <http://intraweb.stockton.edu/eyos/coastal/content/docs/sandy/SouthernMonmouth.pdf> (accessed 13.02.2013).

Schwartz, M.L. (Ed.), 2005. Encyclopedia of Coastal Science. Springer, Dordrecht, The Netherlands.

Seabrook, J., 2013. The beach builders – can the Jersey shore be saved? The New Yorker, 22.

Spoto, M. 2012. Dune Size Determined Extent of Storm Damage on NJ Beaches. NJ.com [Online]. November 18. Available from: <http://www.nj.com/news/index.ssf/2012/11/dune_size_determined_extent_of.html> (accessed 10.01.2014).

United States Army Corps of Engineers National Coastal Mapping Program, 2010. NJ LiDAR Dataset. Available from: <http://www.csc.noaa.gov/dataviewer/> (accessed 10.02.2012).

CHAPTER 6

Observations of the Influence of Regional Beach Dynamics on the Impacts of Storm Waves on the Connecticut Coast During Hurricanes Irene and Sandy

James F. Tait and Ezgi Akpinar Ferrand

ABSTRACT

The Connecticut shoreline is one of the most intensively developed in the country. In many locations, development has relied on the buffering capacity of broad beaches for protection against storms. Much of this development is at risk due to an insufficient understanding of regional beach dynamics. The coast is commonly regarded as "protected" by the presence of Long Island. Nonetheless, Irene and Sandy imposed significant property losses on coastal cities. The most severe damages were due to wave impact in areas with narrow beaches. Small differences (as little as 21 m) in beach width proved to be significant during these storms. Sheltering by Long Island does not prevent coastal erosion during local storms. In the long run, it does prevent the rebuilding of the beach during fair weather by limiting the energy available for shoreward transport. This dynamic makes the beaches naturally erosive and their buffering capacity transient at best.

Keywords: Hurricane Sandy; Hurricane Irene; beach erosion; storm wave impacts; energy asymmetry; Connecticut beach dynamics

6.1 STORM HISTORY

The Connecticut shoreline is one of the most intensively developed shorelines in the country. It has the fifth highest (non-freshwater) coastal population density (980 persons/mile^2) in the United States according to the National Coastal Population Report (National Oceanic and

Learning from the Impacts of Superstorm Sandy. http://dx.doi.org/10.1016/B978-0-12-801520-9.00006-7

Atmospheric Administration, 2013). The total insured value of coastal properties in the state ($567.8 billion in 2012) is the sixth largest among the 16 hurricane-prone Atlantic and Gulf coast states (Air Worldwide, 2013). The value of insured coastal property (defined as insured property in coastal counties) comprises 65% of all insured property in the state, second only to Florida in this respect. The ratio of *value of total insured coastal county property/km of linear shoreline length* for Connecticut is $3.69 billion/km (Air Worldwide, 2013 and Beaver, 2006). This is second only to New York State ($14.3 billion/km). Connecticut coastal development is now at risk, in part because the patterns and policies of development were underinformed by an understanding of regional beach dynamics on this formerly glaciated, fetch-limited shoreline (Figure 6.1).

In Connecticut's experience of storm damage, Hurricane Sandy (October 2012) is closely linked to Hurricane Irene (August 2011) in the sense that people were still in the process of recovering from Irene when Sandy hit. Sandy intensified and extended the amount of damage that had already been done. As an example of how these two storms are convolved,

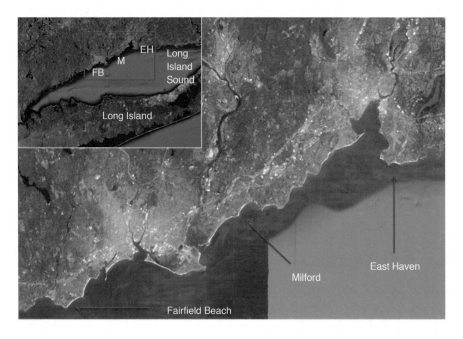

Fig. 6.1. Map and locations of coastal towns that were particularly hard hit by wave damage due to severely eroded beaches. Note the sheltering due to the presence of Long Island. Imagery from Google Earth (9/20/2013).

storm damage records maintained by the East Haven Building Department often cannot attribute damages to a particular storm. Irene hit the Connecticut coast on August 28, 2011 with a storm tide recorded at the NOAA New Haven, CT gage at 2.516 m (msl). The peak surge arrived at 10:36 a.m., coinciding with high tide. Sustained wind speeds were as high as 17.4 m/s with maximum gusts recorded at 22.3 m/s. The wind field was from the south, piling the waters of Long Island Sound directly into the Connecticut coast to the north. As a result of the timing of the surge, the tides and the direction and strength of the wind field, the impacts of Irene on the coast were maximized (Table 1). The Cosey Beach area of East Haven, for example, received national press coverage, losing some 26 houses, according to journalistic reports (for example Hartford Courant, 2011; O'Conner, 2013), along 840 m of beachfront. The most severe damages, not surprisingly, were due to wave impact.

Hurricane Sandy arrived one year later on October 29, 2012. Coastal residents were still attempting to recover from Irene. Although Sandy was a more intense storm generally in terms of maximum wind speeds, diameter, barometric pressure, and maximum storm surge (Fischetti and Mark, 2012), prevailing conditions in Connecticut moderated the storm's impacts. The storm turned west and the eye went into New Jersey. Its forward speed also accelerated to approximately 45 km/h (Ng, 2012). Because of this, the peak storm surge arrived at 8:06 p.m., two hours after a spring low tide (Table 6.1). If not for the acceleration, peak surge

Table 6.1 Comparison of Storm Characteristics Recorded at the New Haven Tide Gage (NOAA)

Storm	Date	Maximum Sustained Wind Velocity (m/s)	Maximum Gusts (m/s)	Peak Surge Magnitude (m) and Time	Tidal Stage at Peak Surge (m)	Maximum Storm Tide Magnitude (m) and Time	Predicted High Tide (m) and Time	Predicted Low Tide (m) and Time
Irene	August 28, 2011	17.4 @ 178	22.3	1.418 @ 10:36 a.m.	1.098	2.516 @ 10:36 a.m.	1.135 @ 11:06 a.m.	−1.083 @ 4:54 a.m.
Sandy	October 29, 2012	10.7 @ 106	25.8	2.784 @ 8:06 p.m.	−0.414	2.721 @ 9:36 p.m.	0.882 @ 11:54 p.m.	−1.026 @ 5:48 p.m.

Note that while Sandy was recognized as the larger and more devastating storm for the U.S. as a whole, Irene arrived in Connecticut with almost the same storm-tide elevation (storm surge plus astronomical tide), had higher sustained wind speeds and had wind blowing seawater directly onshore. All three of these factors had the capacity to intensify wave activity at the shoreline.

would have occurred nearer to a spring high tide. As a result, maximum storm tide was 2.721 m (msl), only 20 cm higher than Irene. Maximum sustained winds reached 10.7 m/s and maximum gusts reached 25.8 m/s. The wind field also played a role in moderating the impacts of Sandy. For most of the day, the wind field was from the northeast, driving water away from the shoreline. Winds shifted to a more easterly direction by 5:00 p.m. and were blowing at an oblique angle to much of the east–west trending Connecticut shore when the peak surge arrived in the New Haven area. Nevertheless, wave heights exceeding 2 m were recorded via buoy in central Long Island Sound (Benson, 2012). These waves did considerable damage upon reaching the shore near the time of peak surge, particularly in the towns of Fairfield Beach, Milford, and East Haven (Figures 6.1 and 6.2). Not only were houses placed at risk, so was infrastructure such as coastal roads.

Fig. 6.2. Examples of storm wave damage from Hurricane Sandy. (A) House in East Haven located at the end of Caroline Road. The beach narrows here and the waves piled sand up on the seaward side of the house then used this sand as a ramp to attack the house. (B) House in Milford located in the Walnut Beach area. The seaward half of the house, which rested on the concrete pad, was demolished. The storm waves eroded the beach out from under the foundation and the house split in half. Photos by James Tait.

6.2 THE INFLUENCE OF BEACH DIMENSIONS ON STORM WAVE IMPACTS

Initial observations of storm wave impact indicate that important controls on the severity of damage were ground elevation, elevation of structures themselves, the presence of seawalls, and the dimensions of the fronting beach. Again, according to initial observations, the most common factor influencing structural damage was the combination of beach width and elevation. In areas where the beaches were particularly narrow, wave damage and loss of structures were at a maximum (Figure 6.3). Damages varied from loss of a porch or deck, to loss of an outer wall, to the complete collapse of the structure. Houses set farther back on the beach profile, on the other hand, endured both Irene and Sandy with only minor damages. This sometimes amounted to no more than deposition of beach sand around the house. For example, of the 43 homes and structures (for example beach club) located on West Silver Sands Beach in East Haven, three homes and a beach club facility suffered serious wave damage (Figure 6.4). The distances from the mean higher high waterline to the respective structures were in the range of 10–12 m. The majority of the structures along the same beach, with distances of 24–40 m from the mean higher high waterline, experienced minor to no wave damage.

On West Silver Sands Beach, many of the homes that escaped serious damage were fairly old and structurally non-robust. These were located fairly far back on the beach profile. On the other hand, many

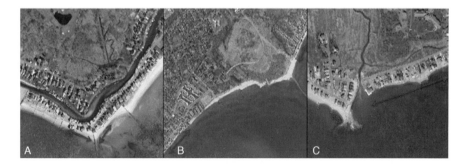

Fig. 6.3. Aerial views of locations where naturally occurring erosion and lack of beach maintenance left coastal homes exposed to wave attack. (A) Fairfield Beach; (B) Milford; and (C) East Haven's Cosey Beach. The red arrows indicate areas with extensive wave damage during both Irene and Sandy. Imagery from U.S. Geological Survey and State of Connecticut, Department of Energy and Environmental Protection.

Fig. 6.4. Aerial view of West Silver Sands Beach in East Haven. Structures too far seaward on the beach profile suffered serious wave damage while older, less robust structures further landward on the beach profile weathered both Irene and Sandy with minimal damage. Imagery from U.S. Geological Survey and State of Connecticut, Department of Energy and Environmental Protection.

of the homes along Cosey Beach Avenue that were farther seaward on the beach profile were seriously damaged despite being newer and built more robustly. For many of these homes, the high tide beach width was essentially zero. In Figure 6.5, for example, the house on the left was undermined and seriously damaged (Figure 6.6). The house on the right suffered the deposition of some sand on its porch and the leakage of seawater into the basement via infiltration through the beach. Both houses are at the same elevation. The principal difference is that the house on the right is located 21 m further back on the beach profile. Most of the wave energy was expended on the beach rather than on the structure itself.

In Figure 6.7, three houses and a beach club occupy different positions on the beach profile. During Irene, the beach club sustained massive damages to beachfront structures. The house marked 44 m

Fig. 6.5. Houses at the end of 1st Avenue in East Haven's Cosey Beach. The house on the left suffered serious wave damage. The house on the right did not. Position of the beach profile made the difference. Imagery from U.S. Geological Survey and State of Connecticut, Department of Energy and Environmental Protection.

Fig. 6.6. Ground view of the house on the left at the end of First Avenue in Figure 6.5. At the time this photo was taken, Sandy had left this house in the intertidal zone. Some of the beach sand was returned when it was removed from the road. The situation is nevertheless problematic. Photo by James Tait.

Fig. 6.7. Houses on Caroline Road in East Haven. Shorefront beach club facilities were severely damaged due to their closeness to the waterline. The house marked 44 m from the waterline is an older cottage with the main floor less than a meter above the sand. It escaped serious damage in both storms due to its position far back on the beach profile. The house marked 26 m from the waterline weathered both Irene and Sandy by virtue of its engineering. It was designed and constructed for the high-energy environment in which it is located. Finally, the house marked 21 m from the waterline is an older cottage that was all but destroyed by the succession of Irene and Sandy. See Figure 6.2 (A) for a close-up view. Imagery from U.S. Geological Survey and State of Connecticut, Department of Energy and Environmental Protection.

from the waterline is not elevated per se but does have an open crawl space surrounded by latticework. It is older and structurally less robust. It survived both Irene and Sandy suffering only some damages to the latticework and deposition of some sand around its exterior. Again wave energy was dissipated across the beach as turbulence and in the physical work of transporting sand landward before reaching the structure. The house marked 26 m from the waterline is elevated and survived both Irene and Sandy. The house is set on pilings that go down 34 m to bedrock. It also has robust storm shutters and the stairs to the deck are removable. Its position on the beach profile, however, is precarious as erosion threatens to undermine the concrete pad beneath the house. The last house on the right, marked 21 m from the shore-line, was devastated by wave attack during Irene and again by Sandy despite having a small seawall. The house is not currently habitable. See Figure 6.2(A) for a close-up.

The width and elevation of the fronting beach are important for protecting houses from waves for two reasons. The energy of a wave is proportional to its height and is expressed as $E = (1/8)\rho g\, H^2$ where E is the energy per unit crest length, ρ is the water density, g is the acceleration of gravity, and H is the wave height in appropriate units. There is a relationship between wave breaking and water depth. This relationship is often expressed as $\gamma_b = H_b/h_b$ where γ_b is a depth-limiting wave breaking criterion determined by the ratio of wave height and water depth. The ratio is often given a value of 0.78 at initial breaking but could be as high as 1.3 (Komar, 1998). After a wave breaks, it begins to dissipate energy as turbulence. The farther it travels across the beach, the more energy it loses. It should be noted, however, that after breaking, it still possesses energy as a translational bore (Komar, 1998). The closer the wave is to its breaking location, the higher the energy level.

During a storm surge, the beach berm can become the surf zone. As such, the difference between the berm elevation and the storm-tide level becomes the effective water depth in the equation above. Storm waves with heights that exceed the water depth will break upon encountering the submerged beach berm. They will then proceed to decay. The more elevated the beach, the smaller the waves that can be supported for a given storm surge. Like an unbroken wave, the turbulent bore also has energy that is proportional to its height. The height of the bore is also depth-limited with γ_b decreasing across the surf zone until it reaches a stable value of 0.42 (Komar, 1998). In other words, the height, and therefore, the energy of the bore can only be about half of the effective water depth.

In sum, higher beach elevations only allow proportionally smaller waves to traverse the beach unbroken with energy intact. Wider beaches provide greater distances over which the energy of broken waves can decay.

6.3 SEASONAL PROFILES AND BEACH EQUILIBRIUM

The Connecticut coast has commonly been thought of as "protected" by the presence of Long Island, New York (for example Lewis, 2014 and Figure 6.1). Despite this protection, Irene and Sandy imposed significant property losses on coastal cities in Connecticut via storm wave impact. This is because the protection against large storm waves that originate in the open Atlantic is counterbalanced by the fact that the presence of Long Island also filters out the fair-weather swell needed to restore sand to the beach.

Common to every oceanography textbook is the notion of seasonal beach profiles, a beach profile being the beach's cross-sectional shape. Originally observed and articulated by Dr. Francis Shepard at Scripps Institute of Oceanography at La Jolla, California, the concept is one in which beaches are in annual equilibrium while seasonally transitioning between a *Winter* or *Storm* condition and a *Summer* or *Fair-weather* condition (Figure 6.8). During winter storms, large waves with short periods erode sand from the subaerial beach and deposit it offshore in the

Fig. 6.8. Boomer Beach in La Jolla, California. (A) Winter of 1940. (B) Summer of 1940. This beach is the "type section" for seasonal beach profiles. Winter storm waves denude the beach but longer period summer swells retrieve the sand from offshore bars and rebuild the beach. Photos by Francis Shepard at Scripps Institute of Oceanography.

form of subaqueous sand bars, causing beaches to be severely cut back. During the fair-weather conditions, the beach is subject to a different wave field consisting of moderately large, long period swell originating from distant storms. This process has been observed innumerable times and is well established in the literature (c.f. Aubrey, 1979; Griggs and Tait, 1988; Komar, 1998; Nordstrom and Inman, 1975; Shepard, 1950; Sonu and Van Beek, 1971; Winant et al., 1975; Ziegler et al., 1959).

Erosion that occurs during the winter is counterbalanced by deposition that occurs during the summer. There is a balance over the year between wave energy that causes the beach to erode, creating offshore bars, and the wave energy that restores the beach to its summer profile with wide berm and steep beach face. Occasionally, a particularly bad storm pulls sand far offshore and into water deep enough that the summer wave field cannot transport the sand back on shore. That sand is permanently lost to the beach and a *disequilibrium* profile prevails until a new equilibrium is established through addition of new sand via littoral drift or bluff erosion, or through readjustment of the remaining beach sediment to the prevailing wave field.

Long Island Sound, the body of water along which Connecticut's shoreline lies, is what is known as a fetch-limited body of water. Waves are created by the wind transferring energy to the water. As this energy transfer occurs over time and distance, the waves grow larger in height. Three factors determine how large the waves will grow: the wind speed, the duration of the wind, and the fetch (or distance) over which the wind blows. The fetch can be determined by the size of a storm or by the dimensions of a body of water. In the case of Long Island Sound, normal fair-weather winds do not have enough distance to create large waves. The waves simply reach the other side of the Sound before they can absorb enough energy to grow large. Because of the presence of Long Island, the only fair-weather wave fields the Connecticut shore is exposed to are generated locally. As such, Long Island Sound is much like a lake in terms of its fair-weather waves. Wave heights at the Connecticut shoreline are typically on the order of centimeters.

The Connecticut coast subsequently experiences a seasonal *wave energy asymmetry*. Storm waves generated by very high winds associated with nor'easters have the requisite energy to transport beach sand offshore, but the fair-weather waves experienced between storms typically lack both the threshold and cumulative energy to rebuild the beach. In a sense, as a fetch-limited shoreline, Connecticut's beaches are in a perpetual state of seasonal disequilibrium. As a consequence, Connecticut's beaches are inherently erosive.

6.4 REGIONAL BEACH DYNAMICS

Nevertheless, Connecticut has had natural beaches since the end of the last ice age and the rise of sea level to its current elevation. Unlike the beaches in many parts of the world, the supply of sand for the beaches is not mainly rivers but the quarrying of glacial drift that is found along much of the Connecticut shoreline in the form of glacial bluffs or headlands that are mantled with glacial sediments (c.f. Otvos, 1965; Sanders and Ellis, 1961). These sediments consist of a mixture of gravel, sand, silts and clays that were eroded from bedrock as the ice sheet associated with the Wisconsin ice sheet advanced over New England. The fine-grained material is transported offshore into deeper waters to become part of the bottom sediments of Long Island Sound or its protected harbors. It may also be carried landward and deposited in local salt marshes. The sand and gravel remain at the shoreline composing the beach (Figure 6.9).

With each storm that erodes sand from the beach, waves may also quarry new sand from the adjacent glacial bluff. In that sense, while

Fig. 6.9. A major source of sand for Connecticut's beaches is the erosion of glacial soils at the shoreline. (A) Sand and gravel eroding from a bluff at Griswold Beach. (B) Sand and gravel at the foot of the bluff in (A). Reworking by the waves has dispersed the finer silt and clay sized grains. Photos by James Tait.

the leading edge of the beach erodes landward under the attack of storm waves at high tide, a beach is maintained from the sand quarried from the retreating glacial bluff. Figure 6.10 shows a beach at Griswold Point near the mouth of the Connecticut River as it appears in 1934 and in 2004. Shoreline retreat has been measured in the past relative to the location of a large rock in the water (see Spiegel, 2013). As the shoreline retreats, the distance between the rock and the glacial bluff at the back of the beach grows. In Figure 6.10, a long-term average retreat can be calculated by dividing the change in distance by the number of years that have elapsed between photos, in this case approximately 0.53 m/yr. The important thing to note is that (allowing for tides) the beach appears to be the same width, albeit in the translated location. This is the regional natural beach dynamic. Shoreline retreat and quarrying of sandy soils maintain the beaches

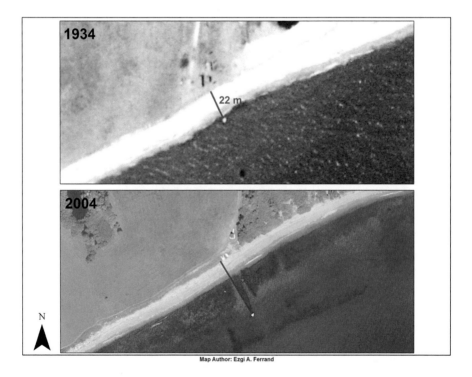

Fig. 6.10. *A well-known rock at Griswold Beach has been used by a number of researchers to estimate erosion rates at this location. In these photos, the glacial bluff seen in Figure 6.8 has eroded 37 m between the two photographs. The beach has been maintained throughout by quarrying of the bluff by waves.* Imagery from the U. S. Geological Survey and mapping by Dr. Ezgi Akpinar Ferrand.

despite the erosional nature of the coast and the asymmetry of erosion and deposition.

6.5 IMPLICATIONS FOR STORM WAVE IMPACTS

The Connecticut coast is very intensively developed. One of the under-appreciated ironies of this is that the development, while in many respects relying on the protection of fronting beaches, has separated the beaches from their supplies of sand in the bluff behind the beach. Particularly problematic is the construction of seawalls for the prevention of coastal retreat. They are common features of much of the Connecticut coast. There are two problems with seawalls in this context. One is that seawalls can have negative impacts on the beaches in front of them by focusing wave energy and inducing scour at the base of the seawall. If the sea-walls interact frequently with large enough waves, erosion of the beach is a predictable result (c.f. Fletcher et al., 1997; Griggs and Tait, 1988; Kraus and McDougal, 1996; Miles et al., 2001; Morton, 1988; Pilkey and Wright, 1988; Plant and Griggs, 1992; and Tait and Griggs, 1991). The larger problem is that these seawalls isolate many Connecticut beaches from their source of sand. On a seasonally asymmetrical coast, this results in the progressive loss of protective beaches. Shoreline construction that depended on the buffering capacity of a wide beach eventually becomes exposed to wave damage. This is Connecticut's coastal dilemma.

To a certain extent, beach nourishment has been used to maintain the beaches. According to Haddad and Pilkey (1998), beach nourish-ment projects have been undertaken at 44 locations along the Connecti-cut coast between 1935 and 1996. Others have been completed in subse-quent years. The Connecticut coast is approximately 160 km long given a linear approximation. This is an average of one beach restoration proj-ect for every 3.6 km of coastline. Many of the projects were undertaken during the 1950s. Today beach nourishment is an expensive proposition. The town of East Haven recently replenished its Town Beach. The beach is only 220 m long and the cost of the project, according to the town engineer's office, was approximately $229,000 and included delivery and grading of 7581 metric tons of sand. As a result, even where beach nour-ishment is an option, initial cost of the project as well as the cost of maintaining the project can become prohibitively expensive and beaches are left to erode.

During Irene and Sandy, the coastal areas that experienced the greatest destruction of houses and other structures were the areas in which the structures were located on highly eroded beaches. Certain stretches of the shoreline in the towns of Fairfield Beach, Milford, and East Haven were particularly hard hit (Figures 6.2 and 6.3). Using the town of East Haven as an example, the Cosey Beach area was developed in the late 1800s as a resort. Over time the beach eroded and seawalls were erected. Predictably the seawalls brought some protection but undoubtedly accelerated beach loss. A classic example of this phenomenon is the loss of the beach at Morris Cove in New Haven. Due to concerns about erosion of the coastal bluff, a seawall was constructed in 1920 (Frances Skelton, Research Librarian Whitney Library, New Haven Historical Museum, personal communications, 2014). Stairways were built into the wall to provide beach access. Now there is no beach and the stairs go down into the water (Figure 6.11).

Figure 6.12 compares the First Avenue area of Cosey Beach in 1934 and 2012. The red rectangle in each photo is geo referenced to the same location. Geo referencing of Figure 6.12 involved using a 2012 National Agriculture Imagery Program (NAIP) orthophoto to assign a spatial location to the known map features in a 1934 tiff raster image. The NAIP imagery is rectified in the UTM coordinate system with a NAD 83

Fig. 6.11. Morris Cove in New Haven, approximately 2 km northwest of Cosey Beach. (A) Hand-colored postcard from 1907 (post-mark). Note the presence of a beach and the erosion of the bluff that supports the houses. (B) Photograph taken from approximately the same spot at low tide in 2014. The seawall was erected to protect the bluff but the beach no longer exists. Postcard by The Hugh C. Leighton, Co., Portland, Maine, USA. Photo by James Tait.

1934   2012

Fig. 6.12. Comparison of the end of First Avenue at Cosey Beach between 1934 and 2012. The red rectangle has been geo referenced to the exact same location in both images. Note that the houses in the rectangle are missing in the 2012 photo and that if they were present, they would be in the water. If the tidal stage on the 1934 image is of a similar stage to that in the 2012 image, then approximately 32 m of beach erosion has occurred during the interval. Imagery from the U. S. Geological Survey and mapping by Dr. Ezgi Akpinar Ferrand.

datum. To geo reference the 1934 image with reference to the 2012 image, ESRI's ArcGIS 10.1 geo referencing toolbar was used.

The tide appears to be at low to mid stage in the 2012 image. If the tidal stage on the 1934 image is of a similar stage, then approximately 32 m of beach erosion has occurred during the 78-year interval. This yields a long-term average retreat rate in excess approximately 0.4 m/yr. If the tidal stage is closer to high in the 1934 image, then the amount and rate of beach erosion is even greater. The houses inside the rectangle do not exist in the 2012 image and would be located in the water if they did. Compare this photo with Figure 6.6. The three houses would have been located seaward and to the west (right) of the house in Figure 6.6.

The west end of Cosey Beach Avenue (Figure 6.13) was the area where most of the homes were lost in East Haven. Many of them were completely demolished. These were among the newest, most robust homes in the area and many of them had seawalls. Their vulnerability stemmed from the lack of a fronting beach. At high tide, the beach is non-existent. And certainly during the storm surges of Irene and Sandy, there was no buffering capacity. Unattenuated waves broke directly against the structures. Figure 6.14 shows wave splash carrying over a two-story home

Fig. 6.13. West end of Cosey Beach Avenue at high tide. There was a broader beach present when these homes were constructed. Seawall-accelerated erosion removed the beach over time. This was the area of maximum damage during Hurricanes Irene and Sandy. Areas of bare soil represent places where houses had been demolished by storm waves. Imagery from Google Earth (9/19/2013).

Fig. 6.14. Wave splash overtops a two-story house in the center of the photo. On the left are the remains of a fairly new house that was destroyed by Hurricane Irene. Photo taken at peak surge during Irene. Photo by James Tait.

during Hurricane Irene at peak surge on Cosey Beach Avenue. To the left of it is a contemporary home that had been completely destroyed by wave impact.

6.6 POLICY IMPLICATIONS

The Connecticut shoreline has been urbanized in such a way that patterns and policies of development have been underinformed by an understanding of regional coastal dynamics. With sea level rise (relative to 1990) expected to reach between 0.5 m and 1.4 m by the end of the century (Rahmstorf, 2007), and with tropical storms predicted to intensify (Knutson et al., 2010), proactive policies are needed. Direct development of the shoreline should be discouraged. Assumptions that the coast is protected by Long Island or that erosion is haphazard rather than systematic need to be rethought. Less expensive alternatives to beach nourishment need to be explored. Much of the eroded sand is lying in the nearshore. Perhaps pumping the sand back onshore rather than trucking it in from outside the region needs to be considered. In the long run, managed retreat from the coastline, particularly in highly vulnerable areas, needs to be explored. It should be noted that a bi-partisan task force (Shoreline Preservation Task Force) formed after Hurricane Irene and authorized by the Connecticut State Legislature has made policy recommendations that, among other things, urge the integration of climate change and sea level rise science into resource and development planning as well as into municipal zoning regulations.

ACKNOWLEDGMENTS

The authors wish to thank Fatima Cecunjanin and Michelle Ritchie for their assistance with Figures 6.1, 6.3, 6.4, 6.5, 6.7, and 6.13. We also wish to thank Megan Coyne for editorial assistance.

REFERENCES

AIR Worldwide Corporation, 2013. The Coastline at Risk: 2013 Update to the Estimated Insured Value of U.S. Coastal Properties. Available from: http://www.air-worldwide.com (accessed 08.01.2014).

Aubrey, D.G., 1979. Seasonal patterns of onshore/offshore sediment movement. Journal of Geophysical Research 84 (C10), 6347–6354.

Beaver, J.C., 2006. U. S. International Borders: Brief Facts Congressional Research Service Report for Congress. Available from: http://fas. org/sgp/crs/misc/RS21729.pdf (accessed 08.01.2014).

Benson, J., 2012. Sound buoys track Sandy's every wind-whipped move. The Day, October 30, 2012. New London, CT. Available from: http://www.theday.com/article/20121030/NWS01/310309949/1/NWShurricanesandy/ (accessed 08.01.2014).

Fletcher, C.H., Mullane, R.A., Richmond, B.M., 1997. Beach loss along armored shorelines on Oahu, Hawaiian Islands. Journal of Coastal Research 13 (1), 209–215.

Fischetti, Mark, 2012. Sandy vs. Katrina, and Irene: Monster Hurricanes By the Numbers. Scientific American. Retrieved from http://www.scientificamerican.com/article/sandy-vs-katrina-and-irene/.

Griggs, G.B., Tait, J.F., 1988. The effect of coastal protection structures on beaches along northern Monterey Bay, California. Journal of Coastal Research 4, 93–111.

Haddad, T.C., Pilkey, O.H., 1998. Summary of the New England beach nourishment experience (1935–1996). Journal of Coastal Research 14 (4), 1395–1404.

Staff Report. Homes destroyed, people missing, and 767,000 without power after Irene. Hartford Courant, 2011. Available from: http://articles.courant.com/2011-08-28/news/hc-hurricane-irene-0829-20110828_1_outage-power-dannel-p-malloy (accessed 08.01.2014).

Knutson, T.R., McBride, J.L., Chan, J., Emanuel, K., Holland, G., Landsea, C., Held, I., Kossin, J.P., Srivastava, A.K., Sugi, M., 2010. Tropical cyclones and climate change. Nature Geoscience 3, 157–163.

Komar, P.D., 1998. Beach Processes and Sedimentation, second ed., Prentice-Hall, Inc., Upper Saddle River, New Jersey.

Kraus, N.C., McDougal, W., 1996. The effects of seawalls on the beach: part 1, an updated literature review. Journal of Coastal Research 12 (3), 691–701.

Lee, G., Nicholls, R.J., Birkemeier, W.A., 1998. Storm-driven variability of the beach-nearshore profile at Duck, North Carolina, USA, 1081–1991. Marine Geology 148 (3–4), 163–177.

Lewis, R.S., 2014. The geology of Long Island Sound. In: Latimer, J.S. et al.,(Ed.), Long Island Sound: Prospects for an Urban Sea. New York, NY, Springer.

Miles, J.R., Russell, P.E., Huntley, D.A., 2001. Field measurements of sediment dynamics in front of a seawall. Journal of Coastal Research 17 (1), 195–206.

Morton, R.A., 1988. The interactions of storms, seawalls, and beaches on the Texas coast. Journal of Coastal Research 4, 113–134.

National Oceanic and Atmospheric Administration, 2013. National Coastal Population Report: Population Trends from 1970 to 2020. Available: http://stateofthecoast.noaa.gov/features/coastal-population-report.pdf (accessed 08.01.2014).

Ng, C., 2012. Hurricane Sandy Arriving Early, to Hit Jersey Shore in the Evening. ABC News, October 29, 2012. Available from: http://abcnews.go.com/US/hurricane-sandy-arriving-early-hit-jersey-shore-evening/story?id=17588404 (accessed 08.01.2014).

Nordstrom, C.E., Inman, D.L., 1975. Sand Level Changes on Torrey Pines Beach, California. U.S. Army Corps of Engineers, Coastal Engineering Research Center, Tech Paper 11-75.

O'Conner, T., 2013. Superstorm Sandy. Connecticut Magazine, January 2013. Available from: http://www.connecticutmag.com/Connecticut-Magazine/January-2013/Story-of-the-Year-Superstorm-Sandy/ (accessed 08.01.2014).

Otvos, Jr., E.J., 1965. Sedimentation-erosion cycles of single tidal periods on Long Island Sound beaches. Journal of Sedimentary Petrology 35 (3), 604–609.

Pilkey, O.H., Wright, III, H., 1988. Seawalls vs. beaches. Journal of Coastal Research 4, 41–64.

Plant, N.G., Griggs, G.B., 1992. Interactions between nearshore processes and beach morphology near a seawall. Journal of Coastal Research 8 (1), 183–200.

Rahmstorf, S., 2007. A semi-empirical approach to projecting future sea-level rise. Science 315, 368.

Sanders, J.E., Ellis, C.W., 1961. Geological and Economic Aspects of Beach Erosion along the Connecticut Coast. Report of Investigations No. 1, State Geological and Natural History Survey of Connecticut.

Sheppard, F.P., 1950. Beach cycles in Southern California. U.S. Army Corps of Engineers, Beach Erosion Board, Tech. Memo 20.

Spiegel, J.E., 2014. New Analysis Pinpoints Change on Connecticut's Long Island Shoreline. The CT Mirror, April 3, 2014. Available: http://ctmirror.org/new-analysis-pinpoints-change-on-connecticuts-long-island-shoreline/ (accessed 08.01.2014).

Sonu, C.J., Van Beek, J.L., 1971. Systematic beach changes on the Outer Banks, North Carolina. Journal of Geology 79, 416–425.

Tait, J.F., Griggs, G.B., 1991. Beach Response to the Presence of a Seawall; Comparison of Field Observations. U. S. Army Corps of Engineer Waterways Experiment Station, Coastal Engineering Research Center, Contract Report CERC-91-1.

Winant, C.D., Inman, D.L., Nordstrom, C.E., 1975. Description of seasonal beach changes using empirical eigenfunctions. Journal of Geophysical Research 80 (15), 1979–1986.

Zeigler, J.M., Hayes, C.R., Tuttle, S.D., 1959. Beach changes during storms on outer Cape Cod, Massachusetts. Journal of Geology 6, 7, 318–336.

CHAPTER 7

Recognizing Past Storm Events in Sediment Cores Based on Comparison to Recent Overwash Sediments Deposited by Superstorm Sandy

J. Bret Bennington and E. Christa Farmer

ABSTRACT

Washover sands deposited by storm surge during Superstorm Sandy along the bay side of Fire Island at Tiana Beach, NY were cored and sampled for grain size analysis. Washover sands at all coring sites exhibited an overall coarsening upward trend, with mean grain sizes varying from 0.25 mm (2 phi) to 0.5 mm (1 phi) from the base to the top of the overwash lobe. Comparison of grain size distributions between washover sands and sands collected from foreshore, berm, backshore, and dune features on the modern ocean beach shows that washover sands are most similar in grain size characteristics to dune sands, particularly near the base of the lobe, but become increasingly similar to beach sands near the top of the lobe. Four sand intervals within a core collected through pre-Sandy marsh deposits at Tiana Beach can be ascribed to overwash from three historical storms (1938 and 1954 hurricanes, 1965 nor'easter) and one unidentified storm based on historical air photographs of the coring site and grain size characteristics. Comparison of grain size characteristics between the historical washover sand layers in core and recent Superstorm Sandy washover sands shows that they share consistent grain size trends that can be used to distinguish washover sands in sediment core from sands deposited by other processes in this and similar barrier island depositional systems.

Keywords: paleotempestology; Hurricane Sandy; overwash; washover; Fire Island

Learning from the Impacts of Superstorm Sandy. http://dx.doi.org/10.1016/B978-0-12-801520-9.00007-9

7.1 INTRODUCTION

Overwash occurs during intense coastal storms when the combination of storm surge and storm waves overtops the beach crest and dune topographic highs of a barrier island and deposits washover sand into back barrier marshes and bays, contributing to the landward migration (rollover) of the barrier island (Donnelly et al., 2009). "Superstorm" (hurricane turned post-tropical cyclone) Sandy made landfall along the New Jersey coast on October 29, 2012, bringing onshore winds and storm surge to bear along the barrier islands of southern Long Island. Strong waves and storm tides in excess of 6 ft above MSL (Blake et al., 2013) resulted in significant overwash along Fire Island, east of the storm's region of maximum impact in northern New Jersey, New York City, and western Long Island. Examination of NOAA aerial images made available on Google Earth shortly after the storm revealed numerous sand lobes deposited by overwash along the length of Fire Island. Particularly good examples of overwash lobes were deposited over the marsh and into Shinnecock Bay in the vicinity of Tiana Beach, west of Shinnecock Inlet on Fire Island (Figure 7.1), a region of the coastline where we have

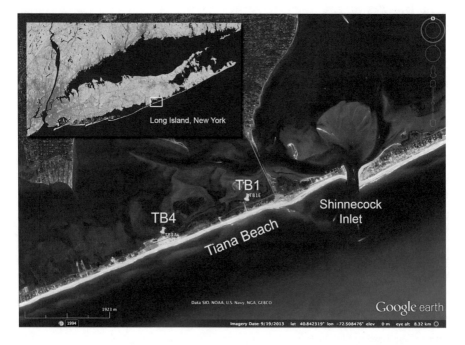

Fig. 7.1. Study localities at Tiana Beach, Long Island, New York plotted on labeled Google Earth image. Inset image of Long Island from NASA Landsat mosaic.

been collecting sediment cores to investigate the history of severe storms impacting Long Island. Thus, Superstorm Sandy presented a unique opportunity to sample and characterize modern washover sands and compare them directly to layers of sand that appear to be washover from past storms in cores collected in close proximity.

Identifying discrete layers of washover in the back barrier sedimentary record and establishing a chronostratigraphic framework for dating washover layers is a widely utilized approach in paleotempestology for developing a chronology of storms dating back before accurate meteorological record keeping (for example McCloskey and Liu, 2013; Scileppi and Donnelly, 2007; Donnelly et al., 2004; Liu and Fearn, 2000) and for assessing the recurrence frequency of major storms like Superstorm Sandy (Liu and Fearn, 2000; Donnelly et al., 2001b, 2004; McCloskey and Keller, 2009; Brandon et al., 2013). Back barrier sediments are dominated by peat and fine-grained layers of mud and silt deposited in the relatively low energy marsh and bay environments. Layers of coarser sediment (shell debris in carbonate environments and quartz sand in clastic environments) are assumed to have been transported across from the ocean side of the barrier by high-energy waves and storm surge, however, other sources of coarse sediment could include wind-blown sand or sand transported into the bay through inlets and breaches in the barrier island and distributed by tidal flow. The objective of this study is to develop additional sedimentological criteria for identifying washover deposits in the sedimentary record by using Superstorm Sandy washover deposits as a modern analog. Other studies have taken a similar approach, studying post-storm washover deposits to better understand storm processes and interpret the sedimentary record of major storm events (Caldwell et al., 2005; Guha et al., 2005; Hawkes and Horton, 2012; Horwitz and Wang, 2005; Leatherman and Williams, 1983; O'Neal-Caldwell and Wang, 2007). In this study, the grain size characteristics of Superstorm Sandy washover deposits were also compared to samples of sand collected from different geomorphological features on the ocean side of the barrier island (foreshore, berm crest, backshore and dune line) to identify potential source areas for the washover sands.

Having identified potential washover deposits in the sedimentary record, a useful resource for attributing washover layers to particular storm events in the recent historical record are aerial photographs taken after major storms to survey the coastline. Although air photos are not available for every storm that impacts the coast, many major storms in the twentieth century were documented in this manner and historic air photos have been used previously to identify overwash events at particular locations along the coast to aid in the dating and interpretation of washover layers in sediment cores (Donnelly et al., 2001b, 2004). In this study, aerial photography was used to identify a particular coring location that was covered by overwash deposits seen in three sets of photos taken after major named storms (the 1938 "Long Island Express" hurricane, Hurricane Carol in 1954, and the Ash Wednesday nor'easter of 1962). This provided examples of known overwash deposits in core that could then be compared directly to the modern overwash deposits provided by Superstorm Sandy.

7.2 STUDY AREA

The study area is located southeast of the town of Hampton Bays, New York on Fire Island, the easternmost barrier island on the south shore of Long Island. In this region of Fire Island the barrier island is relatively narrow (generally less than 0.3 km wide) with a single line of dunes separating the foreshore from the backshore marsh and Shinnecock Bay. A single road, Dune Road, runs the length of the barrier north of the dune line along the southern margin of the high marsh. In the study area, Dune Road commonly floods during spring high tides. Superstorm Sandy overwash deposits were cored in the vicinity of Tiana Beach at locality TB4, approximately 4 km southwest of Shinnecock Inlet. The pre-Sandy sediment core collected for comparison was taken at locality TB1, approximately 1.75 km northeast of TB4 (Figure 7.1). For this study, 5 cores were collected through Superstorm Sandy overwash lobes (Figure 7.2) at sites TB4A (40.829500°, −72.527200°), TB4B (40.829463°, −72.525985°), TB4C (40.829426°, −72.525961°), TB4D (40.830300°, −72.524400°) and TB4E (40.830061°, −72.524331°). A 1.6-m long sediment core was collected from the marsh at site TB1E (40.835900°, −72.504853°) and samples of surface sediment from oceanfront features were collected along the beach approximately 200 m south of TB1E at the end of the beach access road. This unnamed access road crosses the island from the ocean to the bay at locality TB1 and there is a gap in the dune field where it is bisected by the road.

Fig. 7.2. Study locality TB4 before (A) and after (B) Superstorm Sandy. Markers (A–E) indicate sites where cores were collected in overwash lobes during January of 2013.

7.3 METHODS

Sediment cores were collected from overwash lobes by hammering 1-m sections of 3 inch (7.6 cm) aluminum irrigation pipe into the subsurface, capping the pipe, and extracting it using ropes and a farm jack. The longer pre-Sandy marsh sediment core was collected in similar fashion by driving a 2.5 m section of the same aluminum pipe into the surface of the marsh using a vibracore system consisting of a concrete vibrator and custom fabricated connector attached to the pipe. Oceanfront sediment samples were collected by digging a shallow trench with a hand trowel and collecting approximately 1 L of sediment. Duplicate samples were collected from each oceanfront location. Cores were split lengthwise in the laboratory, photographed, and sectioned into 2 cm intervals for grain size analysis. Sediment samples were wet sieved through 250 μm and 64 μm sieves to separate course from fine fractions. After drying, the coarse fraction was sieved through a sieve stack at half phi size intervals (−1 phi to 4 phi) using a Ro-Tap model RX-29 sediment shaker to permit calculation of the weight percent grain size distribution and mean phi size by weight.

Historical aerial photographs used in this study were downloaded at the highest available resolution from the United States Army Corps of Engineers Coastal Hydraulics Laboratory (USACE CHL) website (http://rsm.usace.army.mil/shore/) (US Army Corps of Engineers Coastal Hydraulics Laboratory, 2014). Historical photos were imported into Google Earth as overlay images and referenced to the longitude and latitude grid by scaling and aligning each image to the current Google Earth view of the study area using landmarks such as Dune Road and the drainage canals cut into the marsh for mosquito control prior to the 1930s.

7.4 RESULTS AND INTERPRETATION

At locality TB4, washover deposits from Superstorm Sandy extend as lobate masses from the dune line northward across Dune Road and the marsh (Figure 7.2). At coring sites B, C, D, and E washover sands were deposited directly on the marsh surface, which was most likely flooded at the time of lobe emplacement by high tide. At coring site TB4A the overwash extended into Shinnecock Bay, depositing washover sands into deeper standing water (Figure 7.3(A)). Overwash lobes have distinct distal and lateral margins, although thin deposits of washover sand were observed on the marsh surface beyond the lobe margins (Figure 7.3 (B and C)). Where washover sands were deposited directly on the marsh surface, stalks of marsh grass and some debris items were buried intact and upright (Figure 7.3 (B and D)). No evidence was observed in any of the cores of erosion of the previously deposited sands and peat during the overwash event.

7.4.1 Sediment Analysis of Superstorm Sandy Washover Deposits

Washover deposits observed in core are mostly unstratified except in the lower part of the core at TB4A where thin layers of heavy mineral grains (primarily garnet and magnetite) are seen up to 30 cm above the base

Fig. 7.3. Overwash lobes from Superstorm Sandy at Tiana Beach, New York. (A) Washover sands that extended the shoreline of the barrier island northward into the bay. (B) Distal margin of narrow overwash lobe. (C) Narrow overwash lobe. (D) Marsh grass stalks and debris buried upright in washover sands.

of the lobe in sands washed directly into the bay (Figure 7.4). These thin concentrations of heavy mineral grains are not seen in overwash deposited directly on the marsh and we surmise that they are the result of resuspension of washover sand by waves in the bay as it was being deposited during the overwash event. Grain size analysis shows that the washover sands exhibit a unimodal distribution of predominantly medium sand-sized grains (phi size 1–2). All four cores examined demonstrate a coarsening upward trend from the base to the top of the overwash lobe with basal sands having a mean grain size near 2 phi, coarsening upward to a mean grain size near 1 phi (Figures 7.4 and 7.5).

7.4.2 Comparison of Superstorm Sandy Washover Deposits to Oceanfront Beach Sediments

Weight percent grain size distributions of sediment samples collected from foreshore, berm, backshore, and dune features on the modern ocean beach near coring locality TB1 are shown in Figure 7.6. Beach sands from the foreshore, berm, and backshore have overlapping grain size distributions with mean grain sizes around 1.25 phi. Dune sands are

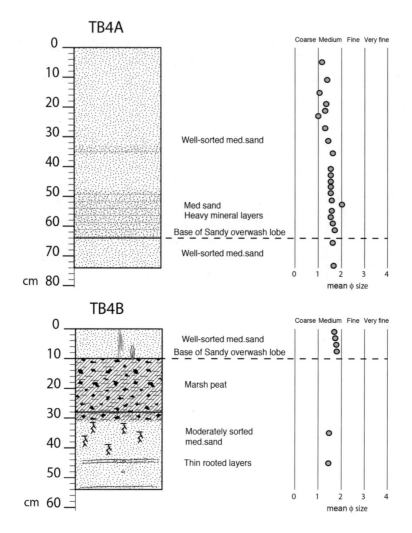

Fig. 7.4. Stratigraphy and mean grain size trends of sediment cores collected through Superstorm Sandy overwash deposits at sampling sites TB4A and TB4B.

significantly finer grained, showing a grain size distribution that is displaced toward finer grains with a mean grain size near 1.5 phi. This is not unexpected given that dune sand is deposited by wind rather than water, which is less competent to transport larger grains due to the low viscosity of air. Representative samples of Superstorm Sandy washover sands taken from near the base of the overwash lobes have grain size distributions that closely match those observed for dune deposits on the ocean beach (although with a slightly smaller mean grain size clustered

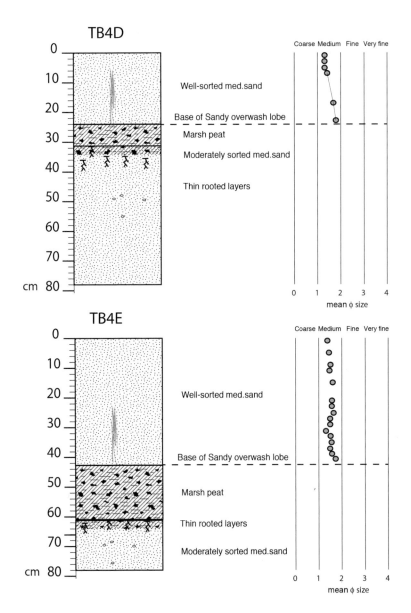

Fig. 7.5. Stratigraphy and mean grain size trends of sediment cores collected through Superstorm Sandy overwash deposits at sampling sites TB4D and TB4E.

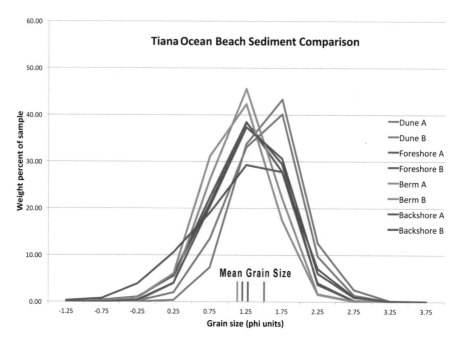

Fig. 7.6. Comparison of grain size distributions and mean grain size of sediment samples collected from different areas of the oceanfront at Tiana Beach. Graph shows replicate samples collected from dune, foreshore, berm, and backshore deposits.

around 1.7 phi – Figure 7.7) suggesting that erosion of the oceanfront dune line was a primary source for washover sands. Comparison of grain size distributions between washover sands from different depths in the overwash lobe at TB4A and average grain size distributions computed for oceanfront beach and dune sands (Figure 7.8) shows a shift from dune-like distributions at the base of the overwash lobe to more beach-like distributions dominated by coarser grain sizes upward to the top of the overwash lobe. This trend can be explained by the overwash event entraining an increasing amount of beach sand after the initial erosion of the dunes.

7.4.3 Comparison of Sandy Washover Deposits to Possible Washover Sands Identified in Core

By superimposing historical air photos of the Tiana Beach region of the barrier island, we were able to identify a specific site (TB1E) that existed within a small indentation of Shinnecock Bay prior to the "Great New England Hurricane" of 1938 and that was subsequently covered by washover sands in 1938, 1954 (Hurricane Carol), and 1962 ("Ash

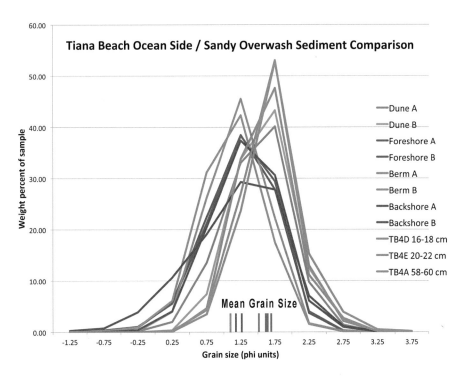

Fig. 7.7. Comparison of grain size distributions and mean grain size of sediment samples collected from ocean beach deposits (red-orange), dune deposits (blue), and Superstorm Sandy overwash deposits (green).

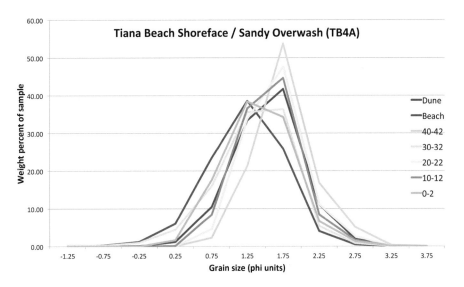

Fig. 7.8. Average grain size distributions of beach (red) and dune (blue) samples compared to samples of washover sands collected from different depths (cm below surface) in the overwash lobe at TB4A. Note that larger depth values indicate samples nearer to the base of the overwash lobe and smaller depth values indicate samples near the surface of the lobe.

Wednesday" nor'easter), with marsh developing on top of the washover between storm events (Figure 7.9). We collected a 1.5-m core at TB1E that reached silty mud at the bottom of the core overlain by four intervals of sand separated by layers of peat and peaty sand (Figure 7.10). We interpret the silty mud at the base of the TB1E core to be a bay bottom deposit and the overlying thick sand layer (142–76 cm) to be the washover from the 1938 hurricane that filled in the southern margin of the bay indent (see Figure 7.9 (A and B)). Thin laminations of heavy

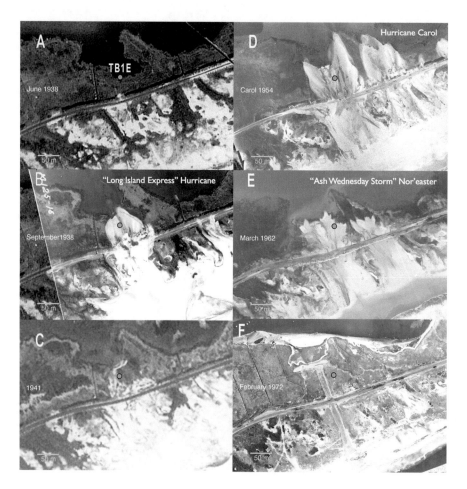

Fig. 7.9. *Historical aerial photos chronicling repeated episodes of washover deposition at core site TB1E. (A) June 1938 showing location of core site at the margin of the bay. (B) September 1938 shortly after the "Long Island Express" or "Great New England Hurricane" of 1938. (C) 1941. (D) 1954 after Hurricane Carol. (E) March 1962 after the "Ash Wednesday Storm" a nor'easter struck on March 6–8, 1962. (F) February, 1972 showing growth of marsh grass on washover sands.*

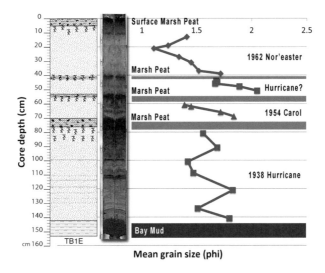

Fig. 7.10. Washover sand intervals in core TB1E identified based on historical aerial photos with mean grain size trends and interpreted stratigraphy.

mineral grains observed in the lower part of this sequence are similar to the layers seen in the Superstorm Sandy washover deposited directly into the bay, corroborating subaqueous deposition. Above the lower sand interval is a 6 cm thick layer of peat (76–70 cm) showing development of salt marsh on top of the washover sands filling in the bay. Although the available air photos conclusively show two additional overwash events following the 1938 hurricane, there are three additional sand intervals separated by peat zones above the post-1938 peat, suggesting three post-1938 overwash events. We are tentatively identifying the sand interval from 70 to 54 cm in the core as washover from Hurricane Carol in 1954 based on the relatively thick 6 cm interval of peat that accumulated prior and the uppermost sand interval from 40 to 6 cm as washover from the nor'easter in 1962 based on the appearance of a thick overwash lobe at the site in the air photo (Figure 7.9(E)). The remaining sand interval from 52 to 42 cm is relatively thin and overlain by a very thin peat interval from 42 cm to 40 cm, suggesting that this sand was deposited shortly prior to the 1964 nor'easter. There are two major storms in the meteorological record that impacted Long Island in the early 1960s that could potentially have produced the unattributed washover deposit, Hurricane Donna, which made landfall as a category 2 storm on eastern Long Island in 1960 (Dunn, 1960) and Hurricane Esther which caused

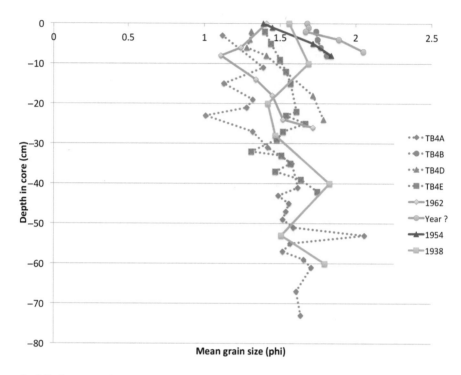

Fig. 7.11. Comparison of mean grain size trends between cores through Superstorm Sandy washover sands (dotted green lines) and sand intervals from core TB1E ascribed to washover sands from historical storms (solid colored lines).

extensive coastal flooding on Long Island in 1961 (National Hurricane Center, 1961). Grain size analysis of the four sand intervals in the TB1E core corroborates their identity as washover sands. Mean grain size in each sand interval exhibits a coarsening upward trend within the medium size range between 1 and 2 phi (Figure 7.10). Direct comparison of Superstorm Sandy washover sediments and inferred washover sediments from TB1E confirms that mean grain size between the two groups exhibits the same range of values and stratigraphic trends (Figure 7.11).

7.5 DISCUSSION AND CONCLUSIONS

Long Island projects almost 200 km oceanward from the middle Atlantic shoreline and is highly vulnerable to impacts from coastal storms such as hurricanes and nor'easters. The barrier islands that mantle the south shore of Long Island are particularly vulnerable to breaching and overwash from intense storms, while at the same time the back barrier marshes and bays are particularly well-situated to record a sedimentary record of intense storm events manifested as discrete layers of washover sand bounded by layers of marsh peat and bay muds (Donnelly et al., 2001a). The sedimentary record of washover deposits from intense coastal storms has been previously studied to develop a history of major storms impacting the coast, augmenting the recent meteorological record for western Long Island (Scileppi and Donnelly, 2007), New Jersey (Donnelly et al., 2001b, 2004), and New England (Donnelly et al., 2001a; Buynevich et al., 2004; Boldt et al., 2010; Fine et al., 2012; Lederer et al., 2000; Madsen et al., 2009; Owen and Donnelly, 2000; Webb and Donnelly, 1999). In these studies, overwash events are identified in core as discrete, unstratified layers of fine to medium sand that abruptly overlie marsh peats or bay muds and that grade into overlying deposits of marsh peat. Superstorm Sandy washover sands exhibit these same sedimentological characteristics, being mostly unstratified, medium sands deposited in abrupt (but non-erosional) contact with bay and marsh sediments. Some thin laminations of heavy mineral grains were observed in washover deposited directly into the bay and we interpret those to be the result of the resuspension and settling of sand by waves coming off the bay during the overwash event. Other studies have noted thin layers and laminations of heavy mineral sands in washover deposited over the eroded supratidal dune field (Wang and Horowitz, 2007; Sedgwick and Davis, 2003), which is clearly not the case here. However, the two circumstances may be analogous, with proximal overwash lobe heavy mineral layers due to resuspension by ocean waves and distal layers due to resuspension by bay waves.

Additional characteristics of washover sands identified from Sandy overwash lobes include the preservation of upright marsh vegetation and a coarsening upward trend in mean grain size (inverse grading) from the base to the top of the overwash lobe, both also reported in similar modern overwash lobes deposited by Hurricane Ike on Galveston and San Luis Islands, Texas (Hawkes and Horton, 2012). Other studies have reported normal grading or absence of grading in washover deposits (see discussion in Sedgwick and Davis, 2003) attributed to deposition under different flow regimes and different amounts of sediment reworking in different depositional circumstances. In this study, we attribute the increase in grain size observed from the base to the top of the overwash lobe to be caused by a shift in sediment source from finer-grained dune sands to coarser beach sands through the duration of the overwash event. This mechanism has been previously proposed to account for some grain size trends in washover deposits (Hawkes and Horton, 2012) and is reasonable, given the massive, unstratified nature of most of the Sandy washover, which demonstrates a lack of the variability in flow regime that would be required to produce graded deposits.

The sedimentological characteristics of overwash-generated deposits described in this study apply to overwash generated from erosion of relatively homogenous sands in the dune field with dune erosion supplying the majority of subsequent washover sediments deposited in marsh and bay margin environments. Previous studies have noted differences between overwash-generated deposits in different regions of the coast with different geomorphic and sedimentological characteristics. For example, O'Neal-Caldwell and Wang (2005) observed significant differences in modern overwash characteristics between the Florida panhandle and Atlantic coasts caused by differences in vegetation and type of sediment. Wang and Horowitz (2007) distinguish between washover deposited through channels resulting from breaching of the barrier and washover deposited by overtopping and erosion of the dune line without breaching and note differences in the sedimentary characteristics of overwash deposited in different back bay environments.

Although the coastal flooding and overwash associated with Superstorm Sandy was extensive, only about 20% of the back barrier marsh in the study area west of Shinnecock Inlet was covered by washover sands. There was no deposition of Sandy washover at coring site TB1E in spite of this location having been covered by washover during four prior storm events within the past century. As other studies have noted (Donnelly et al., 2001a, 2004; Madsen et al., 2009) overwash events are spatially discontinuous, even during intense storms, so that building a comprehensive record of storms using the sedimentary record of washover deposits will require the dating and correlation of washover layers observed in cores collected at multiple sites along the coast.

ACKNOWLEDGMENTS

The authors gratefully acknowledge funding provided by the U.S. Department of Energy (Grant #DE-SC0001985) and several Hofstra University College of Liberal Arts and Sciences Faculty Research and

Development Grants. Many students over several years have assisted with the collecting of cores and the laboratory processing of sediment samples that made this research possible. Undergraduate researchers at Hofstra University have included Jessica Berman, Nika Chery, Emily Dorward, Vanessa Fernandes, Douglas Ferraiolo, Steven Leone, Courtney Melrose, Ashley Persaud, and Dayna Spero. High school summer research interns who have also contributed to this work include Jacob Roday and Sheetal Tolia (Hofstra University Summer Science Research Program 2011) and Brian Zilli, William Berger and Jeremy Silverman (HUSSRP 2013). We thank the Southampton Town Trustees for providing permission to collect sediment cores on town property. Finally, we thank Jeff Donnelly for allowing us to accompany him and his students on a coring expedition in the Great South Bay where we learned some useful tricks of the trade. There is no substitute for experience.

REFERENCES

Blake, Eric S., Kimberlain, Todd B., Berg, Robert J., Cangialosi, John P., Beven II, John L. National Hurricane Center (February 12, 2013) (PDF). Hurricane Sandy: October 22–29, 2012 (Tropical Cyclone Report). United States National Oceanic and Atmospheric Administration's National Weather Service. Retrieved August 05, 2014.

Boldt, K.V., Lane, P., Woodruff, J.D., Donnelly, J.P., 2010. Calibrating a sedimentary record of overwash from Southeastern New England using modeled historic hurricane surges. Marine Geology 275, 127–139, doi:10.1016/j.margeo.2010.05.002.

Brandon, C.M., Woodruff, J.D., Donnelly, J., 2013. How unique was Hurricane Sandy? A comparison of the inundation deposits and surge heights from Hurricane Sandy and the 1821 hurricane. Abstracts with Programs – Geological Society of America 45, 53.

Buynevich, I.V., FitzGerald, D.M., van Heteren, S., 2004. Sedimentary records of intense storms in Holocene barrier sequences, Maine, USA. Marine Geology 210, 135–148, doi:10.1016/j.margeo.2004.05.007.

Caldwell, M.O., Wang, P., Horwitz, M., Kirby, J., Guha, S., 2005. Regional overwash from Hurricanes Frances, Jeanne and Ivan. Transactions of the Gulf Coast Association of Geological Societies 55, 47–56.

Donnelly, C., Hanson, H., Larson, M., 2009. A numerical model of coastal overwash. Proceedings of the Institution of Civil Engineers, Maritime Engineering 162, 105–114, doi:10.1680/maen.2009.162.3.105.

Donnelly, J.P., Bryant, S.S., Butler, J., Dowling, J., Fan, L., Hausmann, N., Newby, P., Shuman, B., Stern, J., Westover, K., Webb, III, T., 2001a. 700 yr sedimentary record of intense hurricane landfalls in southern New England. Geological Society of America Bulletin 113, 714–727.

Donnelly, J.P., Butler, J., Roll, S., Wengren, M., Webb, III, T., 2004. A backbarrier overwash record of intense storms from Brigantine New Jersey. Marine Geology 210, 107–121, doi:10.1016/j.margeo.2004.05.005.

Donnelly, J.P., Roll, S., Wengren, M., Butler, J., Lederer, R., Webb, III, T., 2001b. Sedimentary evidence of intense hurricane strikes from New Jersey. Geology [Boulder] 29, 615–618.

Dunn, Gordon E. (1960). 1960 Monthly Weather Review (PDF). National Hurricane Center. Retrieved 03.02.2008.

Fine, L., Donnelly, J., Martini, A., Woodruff, J., 2012. Calibrating a sedimentary record of hurricane overwash deposition from Quissett Harbor, Woods Hole, MA. Abstracts with Programs – Geological Society of America 44, 67–167.

Guha, S., Horwitz, M.H., O'Neal-Caldwell, M., Wang, P., Kruse, S., 2005. Identifying causes of GPR reflections in hurricane overwash deposits, Santa Rosa Island, Florida. Abstracts with Programs – Geological Society of America 37, 207–1207.

Hawkes, A.D., Horton, B.P., 2012. Sedimentary record of storm deposits from Hurricane Ike, Galveston and San Luis islands, Texas. Geomorphology 171, 172, 180–189.

Horwitz, M., Wang, P., 2005. Sedimentological characteristics and internal architecture of two overwash fans from Hurricanes Ivan and Jeanne. Transactions – Gulf Coast Association of Geological Societies 55, 342–352.

Leatherman, S.P., Williams, A.T., 1983. Vertical sedimentation units in a barrier island washover fan. Earth Surface Processes and Landforms 8, 141–150, doi:10.1002/esp.3290080205.

Lederer, R.W., Cleary, P., Donnelly, J.P., 2000. A stratigraphic record of intense hurricanes in western Connecticut. Abstracts with Programs – Geological Society of America 32, 29–129.

Liu, K., Fearn, M.L., 2000. Reconstruction of Prehistoric Landfall Frequencies of Catastrophic Hurricanes in Northwestern Florida from Lake Sediment Records. Quaternary Research 54, 238–245, doi:10.1006/qres.2000.2166.

Madsen, A.T., Duller, G.A.T., Donnelly, J.P., Roberts, H.M., Wintle, A.G., 2009. A chronology of hurricane landfalls at Little Sippewissett Marsh, Massachusetts, USA, using optical dating. Geomorphology 109, 36–45, doi:10.1016/j.geomorph.2008.08.023.

McCloskey, T.A., Keller, G., 2009. 5000 year sedimentary record of hurricane strikes on the central coast of Belize. Quaternary International 195, 53–68, doi:10.1016/j.quaint.2008.03.003.

McCloskey, T.A., Liu, K.B., 2013. A 7000 year record of paleohurricane activity from a coastal wetland in Belize. Holocene 23, 278–291.

National Hurricane Center (1961). Hurricane Esther Preliminary Report. Retrieved 03.02.2008.

O'Neal-Caldwell, M., Wang, P., 2007. Overwash deposits on Florida barrier islands; a study of contrasting depositional styles. Abstracts with Programs – Geological Society of America 39, 69.

Owen, J., Donnelly, J.P., 2000. Sedimentary evidence of intense hurricane strikes from coastal wetland deposits in western Buzzards Bay, Massachusetts. Abstracts with Programs – Geological Society of America 32, 63–64.

Scileppi, E., Donnelly, J.P., 2007. Sedimentary evidence of hurricane strikes in western Long Island, New York. Geochemistry, Geophysics, Geosystems 8, Q06011, doi:10.1029/2006GC001463.

Sedgwick, P.E., Davis, Jr., R.A., 2003. Stratigraphy of washover deposits in Florida: implications for recognition in the stratigraphic record. Marine Geology 200, 31–48, doi:10.1016/S0025-3227(03)00163-4.

US Army Corps of Engineers Coastal Hydraulics Laboratory, 2014. "Historical Aerial Photos". Historical Aerial Photos. United States Army Corps of Engineers Coastal Hydraulics Laboratory, n.d. Web. 05 August 2014. <http://rsm.usace.army.mil/shore/>.

Wang, P., Horwitz, M.H., 2007. Erosional and depositional characteristics of regional overwash deposits caused by multiple hurricanes. Sedimentology 54, 545–564.

Webb, III, T., Donnelly, J.P., 1999. A 600 year stratigraphic record of intense hurricane landfalls in a Rhode Island salt marsh. Abstracts with Programs – Geological Society of America 31, 77–177.

CHAPTER 8

Trace Metals as a Tool for Chronostratigraphy in Sediment Cores from South Shore Barrier Beach Marshes in Long Island, NY

E. Christa Farmer and J. Bret Bennington

ABSTRACT

Trace metal (Cu, Zn, and Pb) concentration profiles in three sediment cores collected from a saltwater marsh on a barrier beach island near Hampton Bays, NY are consistent with values seen in other Atlantic coast saltwater marshes. Estimation of confidence intervals for the measurements suggests that Pb varies significantly along the depth profiles, while Cu and Zn do not. Peak Pb values in the middle of the profiles are consistent with a higher anthropogenic signal in the middle of the last century. A comparison of grain size characteristics in the sediment cores with grain size characteristics in washover sands known to have been deposited by "Superstorm" (hurricane turned post-tropical cyclone) Sandy reveals which layers in the sediment cores are likely to have been deposited by past storms. The Pb concentrations provide insight into the timing of these storms.

Keywords: sediment cores; sedimentology; paleotempestology; overwash; trace metals; lead; copper; zinc

8.1 INTRODUCTION

The devastating impacts of storms such as "Superstorm" (hurricane turned post-tropical cyclone) Sandy have been well documented in this volume and elsewhere (Rappaport, 2014). In order to prepare effectively for future storms, we need to understand the relative contributions of natural and anthropogenic variability in storm frequency and intensity. This requires characterizing the decadal to centennial patterns in past

Learning from the Impacts of Superstorm Sandy. http://dx.doi.org/10.1016/B978-0-12-801520-9.00008-0

storm behavior. Satellite records are only available for a few decades and are therefore of inadequate length for this task (Goldenberg et al., 2001; Kossin et al., 2014). Turning to the geologic record, however, gives a much longer perspective on storm history. Many studies have exploited various geologic proxies for this goal (Brown et al., 2014; Denommee et al., 2014; Haig et al., 2014; and McCloskey and Liu, 2013, for example, and references therein).

We set out to determine if we could identify hurricane "overwash" layers in sediment cores collected from marshes on the barrier beach islands along Long Island's south shore. Although a number of studies have been published documenting washover deposits from sediment cores in New Jersey (Donnelly et al., 2001b, 2004) and New England (Donnelly et al., 2001a; Buynevich et al., 2004; Boldt et al., 2010; Fine et al., 2012; Lederer et al., 2000; Madsen et al., 2009; Owen and Donnelly, 2000; Webb and Donnelly, 1999), only one study to date, by Scileppi and Donnelly (2007), has reported results from the Long Island region. It is important to sample multiple parts of the coastline to develop an accurate accounting of past storms because of the limited spatial extent of each individual storm, the even smaller section of a storm making landfall that is capable of producing sedimentary features that can be recognized in core section, and the fact that, even in a major event like Superstorm Sandy, only some areas of the coastline experience overwash. Comparing records from only a small number of coastal locations could bias the interpretation of storm history and behavior, because certain storms may not affect the particular locations sampled in one or in a small set of studies. Reynolds-Fleming et al. (2013) also highlight another significant issue in interpretation of sedimentological archives of past storms: can the difference between tropical storms and nor'easters be distinguished?

Establishing the age of various layers in sediment cores is critical for correlating between cores and developing a chronology of past storm events. Besides radiocarbon (^{14}C) dating, trends in pollen concentrations, Cesium (^{137}Cs) and trace metal concentrations, especially copper (Cu), zinc (Zn), and lead (Pb), have all been used to determine the ages of different layers in sediment cores (Boldt et al., 2010; Brown et al., 2014; Donnelly et al., 2001a,b, 2004, 2005; Liu and Fearn, 2000; Scileppi and Donnelly, 2007). None of these other studies, however, incorporate any measure of the sampling error of the trace metal concentrations. Here we investigate the trace metal concentrations in two adjacent sediment cores and their implications for stratigraphy of historical storms in the area, with a special focus on the confidence intervals around these measurements.

8.2 METHODS

Two sediment cores, TB1A and TB1D, were collected using a vibracore system consisting of a concrete vibrator, custom fabricated connector, and aluminum irrigation pipe. TB1A was collected on July 26, 2011 from the Tiana Beach area at 40°50′11.0″N and 72°30′18.9″W, which is approximately 750 m to the southwest of the southern end of the Ponquogue Bridge in Hampton Bays, NY. TB1D was collected on 31 May 2012 from nearly the same location, approximately 10 m to the southeast at 40°50′10.9″N and 72°30′18.73″W (see Figure 8.1).

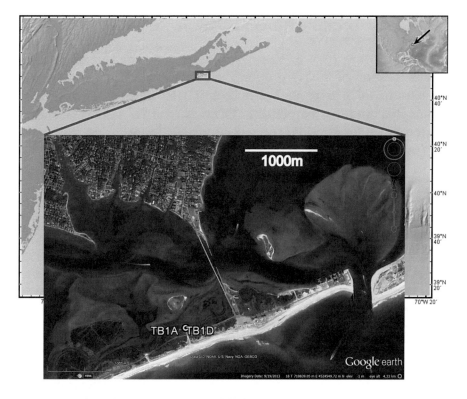

Fig. 8.1. Regional map (from GeoMapApp 3.3.9, available from GeoMapApp.org, Ryan et al., 2009) with inset of the study area (from Google.com, map imagery date 9/19/2003, map data from SIO, NOAA, U.S. Navy, NGA, GEBCO) showing the location where sediment cores TB1A and TB1D were collected. Note Dune Road running nearly east west along the barrier beach island, perpendicular to the Ponquogue Bridge.

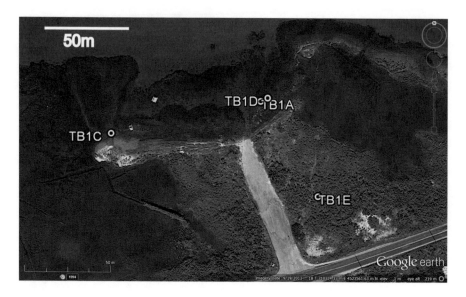

Fig. 8.2. *Small-scale map of the study area (from Google.com, map imagery date 9/19/2003) showing the locations where sediment cores TB1A, TB1C, TB1D, and TB1E were collected. Note Dune Road (paved) running nearly east west at the southern edge of the image.*

The location of TB1D is approximately 60 m to the northeast of the location of a third sediment core, TB1E, which is described in Bennington and Farmer (2014). Cores were split into two halves lengthwise within a few days of collection and photographed that same day (see Figure 8.3B and C).

Samples from the muddier layers of all three cores were prepared for trace metal analysis. Previous efforts have revealed a correlation between sediment grain size and trace metal concentration (Loring, 1991; Yunus et al., 2010), which results in sandier layers generally not containing enough trace metals for analysis (Farmer, unpublished data). Roughly 2 cc of sediment were removed from the interior of the sediment core, from a section that had not been in contact with the metal coring pipe (so as to minimize possible trace metal contamination from that source). These samples were washed through a plastic 150 μm sieve with deionized water. The fine fractions were dried overnight in plastic beakers in a 40 °C oven, weighed, and sent to ALS Minerals in Reno, NV for analysis. Cu, Zn, and Pb concentrations were measured by inductively coupled plasma atomic emission spectrophotometry (ICP-AES). A four-acid digestion of the sample was used to liberate the trace metals for ICP-AES analysis.

Table 8.1 Sieve Sizes Used in Grain Size Analysis		
(A) Range (Φ)	(B) Range (mm)	(C) Midpoint (Φ)
−1.5 to −1	2.83 to 2	−1.25
−1 to −0.5	2 to 1.41	−0.75
−0.5 to −0	1.41 to 1	−0.25
0 to 0.5	1 to 0.71	0.25
0.5 to 1.0	0.71 to 0.50	0.75
1.0 to 1.5	0.50 to 0.35	1.25
1.5 to 2.0	0.35 to 0.25	1.75
2.0 to 2.5	0.25 to 0.177	2.25
2.5 to 3.0	0.177 to 0.125	2.75
3.0 to 3.5	0.125 to 0.088	3.25
3.5 to 4.0	0.088 to 0.063	3.75

Grain size was measured in sandier layers using a Ro-Tap model RX-29 sediment shaker and a series of sieves (see Table 8.1).

Samples were shaken for 15 min with 150 ± 10 taps per minute and 278 ± 10 oscillations per minute.

8.3 RESULTS

Cu concentrations ranged from 6 to 28 ppm in TB1A, 14 to 38 ppm in TB1D, and 8 to 43 ppm in TB1E; Zn concentrations ranged from 32 to 80 ppm in TB1A, 54 to 152 ppm in TB1D, and 35 to 75 ppm in TB1E; and Pb concentrations ranged from 13 to 140 ppm in TB1A, 48 to 105 ppm in TB1D, and 15 to 115 ppm in TB1E (see Table 8.2 and Figures 8.3 and 8.4).

Grain size measurements are displayed in Table 8.3 and Figures 8.5 and 8.6: each size class is reported as a percentage by weight of the total weight of the >63 micron fraction of the sample at that depth.

8.4 DISCUSSION

8.4.1 Depth Adjustment for Disturbance

During the core retrieval process, a gap opened up between sediment layers in TB1A; empty space and some disturbed sediment can be seen in the core between 17 and 50 cm (see Figure 8.3B). To compensate for this distur-bance of the lower sediment layers, we subtracted 33 cm from the depth of

Table 8.2 Trace Metal Concentrations Measured in TB1A and TB1D				
Core	Depth (cm)	Cu (ppm)	Zn (ppm)	Pb (ppm)
TB1A	10–11	17	64	32
TB1A	57–58	28	80	140
TB1A	134–135	6	32	17
TB1A	142–143	8	35	13
TB1D	2–3	23	72	55
TB1D	10–11	29	152	92
TB1D	18.5–19.5	14	54	105
TB1D	30.5–31.5	38	81	52
TB1D	70–71	26	70	48
TB1E	1–2	34	67	49
TB1E	40–41	43	75	100
TB1E	54–55	15	63	64
TB1E	70–71	21	48	88
TB1E	74–75	17	41	115
TB1E	142–143	8	48	19
TB1E	150–151	8	35	15

the measurements made below 17 cm. This effectively raises the lower measurements in Figure 4A up to the depth where they would have been, had the gap not opened up and displaced the lower sediment layers.

8.4.2 Trace Metal Concentrations

We estimate confidence intervals for the trace metal concentration measurements from analysis of a surface grid of 24 locations in a nearby marsh. In those grid samples, the standard deviation of Cu measurements was 8 ppm, or 31% of the average measured value; the standard deviation of Zn measurements was 12 ppm, or 14% of the average measured value; and the standard deviation of Pb measurements was 6 ppm, or 14% of the average measured value (Farmer and Silverman, unpublished data). These confidence intervals are plotted in ppm as the horizontal error bars in Figures 8.3 and 8.4.

Given these estimates of confidence intervals for these data, the Cu concentrations in TB1A and TB1D hardly vary from one depth to the next. In TB1A in particular, the difference between the minimum Cu value (6 ppm) to the maximum Cu value (28 ppm) is only 22 ppm. For

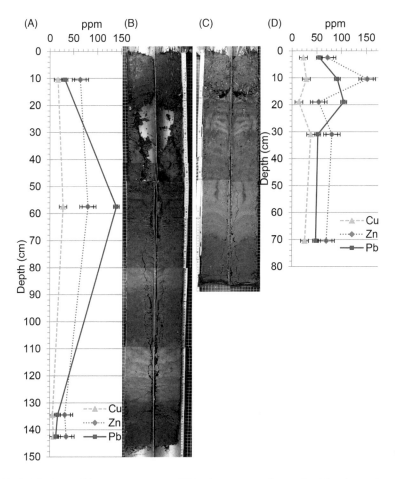

Fig. 8.3. (A) Cu, Zn, and Pb concentrations in core TB1A, (B) composite photographs of the split core TB1A, (C) composite photographs of the split core TB1D, and (D) Cu, Zn, and Pb concentrations in core TB1D. Note all data are shown on the same depth scale.

comparison, the standard deviation of ±8 ppm gives a confidence interval range of 16 ppm. The situation in TB1D is hardly different: the difference between the minimum Cu value (14 ppm) and the maximum Cu value (38 ppm) is only 24 ppm. There is only a slightly larger range in Cu concentrations in TB1E of 35 ppm. Thus, we believe that the subsurface Cu variations are not significant and do not contribute any useful information to our analysis.

The Zn concentrations vary more widely. A different attempt that we made to estimate confidence intervals on these data, however, suggests that

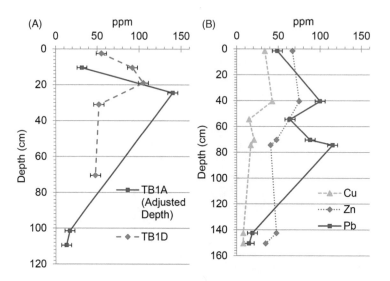

Fig. 8.4. (A) Pb (ppm) in core TB1A (on the adjusted depth scale) and core TB1D, (B) Cu (ppm), Zn (ppm), and Pb (ppm) in TB1E. Note that the depth scales in the two panels have been adjusted in order to align the 101.5 cm sample in TB1A with the 142.5 cm sample in TB1E (see text for discussion).

Table 8.3 Sediment Grain Sizes Measured in Various Depths of TB1D

φ	φ midpoint											φ	Mean (φ)	Mean (mm)
<-1	-0.75	-0.25	0.25	0.75	1.25	1.75	2.25	2.75	3.25	3.75	>4.0			
mm:	mm midpoint													
>2	1.68	1.19	0.84	0.59	0.42	0.30	0.21	0.15	0.11	0.07	<0.063			
Depth (cm)														
1–2	2.73	2.34	0.39	9.38	5.47	7.03	16.02	22.66	16.80	12.11	5.08	36.46	1.94	0.26
10–11	19.24	4.21	5.23	9.80	12.35	8.50	12.49	3.78	9.66	8.21	6.54	19.62	1.02	0.49
17–18	71.25	1.38	2.30	4.07	5.32	5.06	4.77	3.06	1.61	0.76	0.43	0.82	-0.56	1.47
18–19	8.68	5.33	7.71	13.05	15.68	13.30	12.85	10.48	7.07	3.53	2.31	4.92	1.01	0.50
19–20	1.03	0.78	0.98	1.66	2.67	3.07	3.64	3.51	4.93	3.44	1.74	9.32	1.81	0.29
25–26	0.17	0.12	0.29	1.56	8.92	25.61	37.15	15.65	6.30	2.61	1.62	3.14	1.71	0.31
30–31	0.16	0.12	0.40	1.17	5.98	18.57	39.93	22.46	7.86	2.38	0.97	2.56	1.81	0.29
35–36	1.03	0.89	1.88	4.73	12.87	25.52	34.14	14.09	4.03	0.60	0.21	1.02	1.46	0.36
86–87	0.00	0.08	0.28	1.10	5.82	20.97	44.06	22.58	4.45	0.43	0.24	1.01	1.73	0.30

Each size class is reported as a percentage by weight of the total weight of the >63 micron fraction of the sample at that depth.

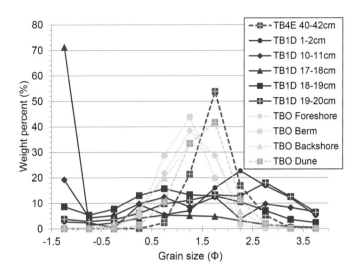

Fig. 8.5. Comparison of grain size of upper (<20 cm depth) TB1D sediments with Superstorm Sandy deposits (TB4E) and oceanfront samples.

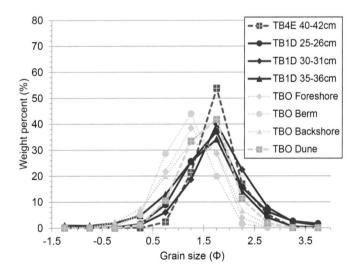

Fig. 8.6. Comparison of grain size of lower (>20 cm depth) TB1D sediments with Superstorm Sandy deposits (TB4E) and oceanfront samples.

the Zn data do not contribute useful information to this study either. We measured triplicate samples at the same depth (109–110 cm) in another core: TB1C, located at 40°50′10.4″N, 72°30′22.6″W (see Figure 8.2), which is approximately 80 m to the southwest of TB1A. The standard deviation of Cu and Pb for those triplicate samples was smaller than the standard deviation found in the grid samples described above, and are not plotted in Figure 8.3A and D. The standard deviation of Zn in those triplicate samples, however, was found to be 17 ppm, or 38% of the average measured value, which is much larger than the standard deviation in the grid samples (Farmer, unpublished data), and are plotted as the larger horizontal error bars in Figure 8.3A and D. Comparing the difference between the minimum and maximum Zn values in TB1A (32 ppm) and in TB1E (40 ppm) with twice the standard deviation of these triplicate values (34 ppm) suggests that the Zn concentrations measured in these two cores do not vary significantly. The range in Zn values found in TB1D is larger (98 ppm). The Zn concentration measured in the 10–11 cm sample is almost twice the next largest value, however. If this sample is discounted as an outlier, the range in TB1D is only 27 ppm. This is more comparable to the Zn concentration range in the other two sediment cores, and also the same magnitude as our estimate of the confidence interval: this all suggests that subsurface Zn concentrations do not contribute anything useful to our analysis either.

The Pb concentrations, however, covered a range of 127 ppm in TB1A, 57 ppm in TB1D, and 100 ppm in TB1E. Our estimate of the confidence interval for Pb concentration measurements in these samples of ±6 ppm suggests that these variations are significant. Once plotted on an adjusted depth scale (see Section 8.4.1), the trends in cores TB1A and TB1D are similar (see Figure 8.4A): concentrations are relatively low in the lower sections of the core, rise to a peak between 19 and 25 cm, then decline again in the upper layers of the core. Pb concentrations in TB1E follow a broadly similar trend of low concentrations at the lower end of the core, a rise to highest concentrations in the middle depths, and a decrease to lower concentrations again in the upper level of the core (see Figure 8.4B). In the middle of the TB1E core, however, between the highest two concentrations at 74.5 and 40.5 cm, the values decrease almost to the level seen in the top layer of the core. The confidence intervals on these measurements, as described above, suggest that this dip is significant: see discussion in Section 8.4.4.

What do other studies of trace metal concentrations in nearby salt marshes suggest about our results? Bricker (1993) measured trace metal concentrations in sediment cores from a Rhode Island salt marsh and found a similar subsurface trend in Pb concentrations as that found in the TB1A and TB1D cores. Chronology of the marsh cores was established independently with excess ^{210}Pb, and indicated that peak Pb concentrations were reached in the sediments in the 1950s (Bricker, 1993). Cochran et al. (1998) measured trace metal concentrations in several sediment cores from salt marshes on the North shore of Long Island, and similarly dated those layers using excess ^{210}Pb. They found peak Pb concentrations around 1970 at sites on western Long Island, and around 1980 at sites on eastern Long Island (Cochran et al., 1998). Kim et al. (2004) also found a similar subsurface trend in Pb concentrations in a sediment core from a Delaware salt marsh, with peak values in the late 1940s or early 1950s (Kim et al., 2004). These studies suggest that our peak Pb concentrations between 19 and 25 cm in TB1A and TB1D represent a time interval between 1950 and 1980. This is consistent with the phase-out of leaded gasoline in 1995 and subsequent reductions in atmospheric sources of Pb pollution (Datko-Williams et al., 2014).

How do the values of Pb concentrations that we found in the sediment cores from the Hampton Bays area compare to values measured in other salt marshes? The peak Pb concentrations found in TB1A (140 ppm at 24–25 cm adjusted depth), TB1D (105 ppm at 18.5–19.5 cm depth), and TB1E (115 ppm at 74–75 cm depth) are all similar in magnitude to the peak values of 100–250 ppm found between 10 and 30 cm in most of the western Long Island barrier beach island salt marsh cores studied by Scileppi and Donnelly (2007). Their easternmost core, which of all their cores was nearest to a road, had a peak of around 800 ppm. Bricker found higher concentrations, with peak values between 300 and 600 ppm (Bricker, 1993). Cochran et al. (1998) found even higher concentrations on western Long Island, with values over 1000 ppm in an Alley Pond site. Peak concentrations on central Long Island sites were lower, closer to 300 ppm, and even lower on eastern Long Island sites. The authors explained this gradient as a decrease in the distance from major roads and the influence of New York City to the west (Cochran et al., 1998). The peak TB1A, TB1D, and TB1E Pb concentrations are higher than the surface values found in the "forested" and "suburban" settings in Sanger et al. (1999), but comparable in magnitude to those found in the setting they labeled "industrial" (17.9–109.9 ppm). Our surface samples are comparable in value to those found in their "forested" and "suburban" settings.

Reis et al. (2009) designated a salt marsh in the Minho estuary in Portugal as a "pristine area" because they measured Pb concentrations largely below the levels considered to have negative effects in organisms. They used the "Effects Range Low" value of 47 ppm for Pb found by Long et al. (1995) as their threshold. Yunus et al. (2010) measured Pb concentrations in surface sediments from a Malaysian estuary of 6.80–86.49 ppm, with an average of 29.98 ppm. They calculated "enrichment factors" based on a comparison of measured values with concentrations found in the Earth's crust, and concluded that the Pb in the area was "predominantly anthropogenic in origin" (Yunus et al., 2010). These studies suggest that the Hampton Bays area was experiencing significant anthropogenic input during the middle of the last century, given that the peak values are comparable to those classified as "anthropogenic" in the Yunus et al. (2010) study. Pb inputs in the Hampton Bays area today are relatively low, compared to the threshold for harm to organisms of 47 ppm established by Long et al. (1995).

8.4.3 Grain Size Data

Comparing the grain size characteristics of the sediment in the TB1D core with the characteristics of washover sediment that was deposited by Superstorm Sandy further illuminates the environmental history of the TB1A and TB1D location. Figure 8.5 shows the grain size distribution (in weight percent) of several samples from the upper depths of TB1D (above 20 cm), while Figure 8.6 shows samples from lower depths of TB1D (below 20 cm). Both panels also show the grain size distributions of samples collected from nearby oceanfront locations (TBO foreshore, berm, backshore, and dune) as well as the grain size distribution of a sample taken from the base of a layer of Superstorm Sandy washover sediment (core TB4E). See Bennington and Farmer (2014, this volume) for a description of the collection of Sandy washover and beachfront sediments.

The grain size distributions of upper (above 20 cm) TB1D sediment layers are not very similar to any of the beachfront samples or the Sandy washover layer (see Figure 8.5). The upper layers of TB1D are quite coarse, and fairly evenly distributed across the grain size categories (other than the category which includes all particles larger than -1 Φ). The lower layers of TB1D, however, are much more similar to the dune sample and Sandy washover in their grain size characteristics (see Figure 8.6), with a peak grain size of 1.75 Φ and a much more narrow distribution overall.

8.4.4 Implications for Stratigraphy

The aerial photographs discussed in Bennington and Farmer (2014, in this volume, see Figure 7.9) imply that the lowermost layers of all three sediment cores (TB1A, TB1D, and TB1E) should be bay bottom. Perhaps the darker layers below 71 cm in TB1D (see Figure 8.3B) were deposited in this environment. It is possible that these layers contain a higher percentage of organic material, although they do not contain a higher percentage of smaller grain size material (the percentage of material greater than 4 Φ is actually lower in the deepest two layers of the core than in most of the other depths; see Table 8.3). The low Pb concentrations and grain size distributions most similar to the Sandy overwash suggest that the sandy layers between 71 cm and 20 cm in TB1D are overwash deposits. If the peak Pb concentration at 19 cm in TB1D represents the 1970s, this timing would be consistent with deposition of the sandy layers between 71 and 20 cm during either Hurricane Carol in 1954 (National Weather Service 2005) or the nor'easter in 1962 (Morton et al., 2014), as the aerial photographs suggest. This timing would also be consistent with the very coarse material at 19 cm in TB1D being deposited during the construction of a short road spur north from Dune road. This spur road appears to be absent in the 1962 aerial photo (see Figure 7.9E) but can be clearly seen in the 1972 aerial photo (Figure 7.9F).

The pattern in Pb concentrations in TB1E is particularly interesting because of the dip in values between peaks at 40.5 cm and 74.5 cm. Is this double peak due to remobilization of Pb after the sediment has accumulated? Cochran et al. (1998) note in their study of sediment cores from the north shore of Long Island that "The preservation of distinctly different profiles of ^{210}Pb and Pb in the marshes sampled [Figures 5(b) and 6] strongly suggests that these metals are not being diagenetically mobilized." This suggests that remobilization is unlikely at our site, as well. Because we were only able to measure trace metal concentrations in muddy layers, and TB1A and TB1D encompass fewer such layers than TB1E, the profiles of Pb in all three cores are consistent: TB1A and TB1D are likely simply missing the archive of Pb concentrations that would show the secondary peak seen in TB1E. Kim et al. (2004) also observed a secondary peak in Pb concentrations in their study of a Delaware salt marsh, attributing the later one to a local (non-gasoline) pollution source, perhaps steel production. Cochran et al. (1998) observed a secondary peak in Pb concentrations at their Nassau County site (Caumsett Park), but did not find secondary peaks at either of their Suffolk County sites further to the northeast. Perhaps prevailing wind directions were such that industrial sources in Nassau County account for the secondary peak at our site?

When compared with the historical aerial photographs (as discussed in Bennington and Farmer, this volume), the pattern of Pb concentrations in TB1E further clarifies the stratigraphy in the area. Because the historic peak in atmospheric Pb pollution in the eastern U.S. is generally assigned to some period between the 1950s and 1970s (Cochran et al., 1998; Kim et al., 2004; see also Section 8.4.2 above), the Pb peak at 74.5 cm is likely from sediment deposited during that time period. This is consistent with the inference that the sandy layer between 55 and 70 cm was deposited by Hurricane Carol in 1954 rather than the 1944 hurricane.

8.5 CONCLUSIONS

A comparison of grain size data and trace metal concentrations from three sediment cores with historical aerial photographs of the area provides a coherent chronology of major storms in the back bay sedimentary record on Fire Island near Hampton Bays, NY. Although no significant trends were observed in copper and zinc concentrations in the study cores, trends in lead are consistent with those observed in other studies and can be used to correlate overwash events occurring in the 1950–1970 time range.

Grain size data from Superstorm Sandy overwash deposits are useful for confirming the identity of washover sands in sediment cores and for ruling out an overwash origin for some sand layers observed in core.

No sedimentological characteristics were observed that would distinguish between washover deposited by tropical storms (or hurricanes) and washover from extra-tropical storms (nor'easters) in the sediment record.

ACKNOWLEDGMENTS

The authors gratefully acknowledge funding provided by the U.S. Department of Energy (Grant#DE-SC0001985) and several Hofstra University College of Liberal Arts and Sciences Faculty Research and Development Grants. We could not have performed this research without sediment coring and sample preparation assistance by Hofstra University undergraduate students Steven Leone, Tamunoisoala LongJohn, Ashley Persaud, Nika Chery, Dayna Spero, Emma Kast, and Courtney Melrose; Jacob Roday and Sheetal Tolia (Hofstra University Summer Science Research Program 2011); Brian Zilli, William Berger and Jeremy Silverman (HUSSRP 2013); Ben Senzer and Sneh Shah (HUSSRP 2014); and students in Dr. Farmer's Fall 2013 Paleoclimatology course. We also appreciated having permission from the Southampton Town Trustees to collect sediment cores on their property.

REFERENCES

Bennington, JB., Farmer, E.C., 2014. Recognizing past storm events in sediment cores based on comparison to recent overwash sediments deposited by Superstorm Sandy. Bennington and Farmer, (Ed.), Learning from the Impacts of Superstorm Sandy, Elsevier (this volume).

Boldt, K.V., Lane, P., Woodruff, J.D., Donnelly, J.P., 2010. Calibrating a sedimentary record of overwash from Southeastern New England using modeled historic hurricane surges. Marine Geology 275, 127–139, doi:10.1016/j.margeo.2010.05.002.

Bricker, S.B., 1993. The history of Cu, Pb, and Zn inputs to Narragansett Bay Rhode-island as recorded by salt-marsh sediments. Estuaries 16, 589–607.

Brown, A.L., Reinhardt, E.G., Van Hengstum, P.J., Pilarczyk, J.E., 2014. A coastal Yucatan sinkhole records intense hurricane events. Journal of Coastal Research 30, 418–428.

Buynevich, I.V., FitzGerald, D.M., van Heteren, S., 2004. Sedimentary records of intense storms in Holocene barrier sequences, Maine, USA. Marine Geology 210, 135–148, doi:10.1016/j.margeo.2004.05.007.

Cochran, J.K., Hirschberg, D.J., Wang, J., Dere, C., 1998. Atmospheric deposition of metals to coastal waters (Long Island Sound, New York, USA): evidence from saltmarsh deposits. Estuarine Coastal and Shelf Science 46, 503–522.

Datko-Williams, L., Wilkie, A., Richmond-Bryant, J., 2014. Analysis of U.S. soil lead (Pb) studies from 1970 to 2012. Science of the Total Environment 468, 854–863.

Denommee, K.C., Bentley, S.J., Droxler, A.W., 2014. Climatic controls on hurricane patterns: a 1200-y near-annual record from Lighthouse Reef, Belize. Scientific Reports, 4.

Donnelly, J.P., Bryant, S.S., Butler, J., Dowling, J., Fan, L., Hausmann, N., Newby, P., Shuman, B., Stern, J., Westover, K., Webb, III, T., 2001a. 700 yr sedimentary record of intense hurricane landfalls in southern New England. Geological Society of America Bulletin 113, 714–727.

Donnelly, J.P., Roll, S., Wengren, M., Butler, J., Lederer, R., Webb, III, T., 2001b. Sedimentary evidence of intense hurricane strikes from New Jersey. Geology [Boulder] 29, 615–618.

Donnelly, J.P., Butler, J., Roll, S., Wengren, M., Webb, III, T., 2004. A backbarrier overwash record of intense storms from Brigantine, New Jersey. Marine Geology 210, 107–121, doi:10.1016/j.margeo.2004.05.005.

Donnelly, J.P., 2005. Evidence of past intense tropical cyclones from backbarrier salt pond sediments: a case study from Isla de Culebrita, Puerto Rico, USA. Journal of Coastal Research, 201–210.

Fine, L., Donnelly, J., Martini, A., Woodruff, J., 2012. Calibrating a sedimentary record of hurricane overwash deposition from Quissett Harbor, Woods Hole, MA. Abstracts with Programs – Geological Society of America 44, 67–167.

Goldenberg, S.B., Landsea, C.W., Mestas-Nunez, A.M., Gray, W.M., 2001. The recent increase in Atlantic hurricane activity: causes and implications. Science 293, 474–479.

Haig, J., Nott, J., Reichart, G.J., 2014. Australian tropical cyclone activity lower than at any time over the past 550–1,500 years. Nature 505, 667-+.

Kim, G., Alleman, L.Y., Church, T.M., 2004. Accumulation records of radionuclides and trace metals in two contrasting Delaware salt marshes. Marine Chemistry 87, 87–96.

Kossin, J.P., Emanuel, K.A., Vecchi, G.A., 2014. The poleward migration of the location of tropical cyclone maximum intensity. Nature 509, 349.

Lederer, R.W., Cleary, P., Donnelly, J.P., 2000. A stratigraphic record of intense hurricanes in western Connecticut. Abstracts with Programs – Geological Society of America 32, 29–129.

Liu, K.B., Fearn, M.L., 2000. Reconstruction of prehistoric landfall frequencies of catastrophic hurricanes in northwestern Florida from lake sediment records. Quaternary Research 54, 238–245.

Long, E.R., Macdonald, D.D., Smith, S.L., Calder, F.D., 1995. Incidence of adverse biological effects within ranges of chemical concentrations in marine and estuarine sediments. Environmental Management 19, 81–97.

Loring, D.H., 1991. Normalization of heavy-metal data from estuarine and coastal sediments. ICES Journal of Marine Science 48, 101–115.

Madsen, A.T., Duller, G.A.T., Donnelly, J.P., Roberts, H.M., Wintle, A.G., 2009. A chronology of hurricane landfalls at Little Sippewissett Marsh, Massachusetts, USA, using optical dating. Geomorphology 109, 36–45, doi:10.1016/j.geomorph.2008.08.023.

McCloskey, T.A., Liu, K.B., 2013. A 7000 year record of paleohurricane activity from a coastal wetland in Belize. Holocene 23, 278–291.

National Weather Service, 2005. Hurricane Carol [Online]. Place Published: Taunton, MA. Available from: http://www.erh.noaa.gov/box/hurricane/hurricaneCarol.shtml (accessed 24.05.2014).

Morton, R., Guy, K., Hill, H. 2014. Morphological Impacts of the March 1962 Storm on Barrier Islands of the Middle Atlantic States. Available from: http://coastal.er.usgs.gov/hurricanes/historical-storms/march1962/ (accessed 24.05.2014).

Owen, J., Donnelly, J.P., 2000. Sedimentary evidence of intense hurricane strikes from coastal wetland deposits in western Buzzards Bay, Massachusetts. Abstracts with Programs – Geological Society of America 32, 63–64.

Rappaport, E.N., 2014. Fatalities in the United States from Atlantic Tropical Cyclones. Bulletin of the American Meteorological Society 95, 341–346.

Reis, P.A., Antunes, J.C., Almeida, C.M.R., 2009. Metal levels in sediments from the Minho estuary salt marsh: a metal clean area? Environmental Monitoring and Assessment 159, 191–205.

Ryan, W.B.F., Carbotte, S.M., Coplan, J.O., O'hara, S., Melkonian, A., Arko, R., Weissel, R.A., Ferrini, V., Goodwillie, A., Nitsche, F., Bonczkowski, J., Zemsky, R., 2009. Global multi-resolution topography synthesis. Geochemistry Geophysics Geosystems, 10.

Sanger, D.M., Holland, A.E., Scott, G.I., 1999. Tidal creek and salt marsh sediments in South Carolina coastal estuaries: I. Distribution of trace metals. Archives of Environmental Contamination and Toxicology 37, 445–457.

Scileppi, E., Donnelly, J.P., 2007. Sedimentary evidence of hurricane strikes in western Long Island, New York. Geochemistry Geophysics Geosystems 8, Q06011.

Webb, III, T., Donnelly, J.P., 1999. A 600 year stratigraphic record of intense hurricane landfalls in a Rhode Island salt marsh. Abstracts with Programs – Geological Society of America 31, 77–177.

Yunus, K., Ahmad, S.W., Chuan, O.M., Bidai, J., 2010. Spatial distribution of lead and copper in the bottom sediments of Pahang River Estuary, Pahang, Malaysia. Sains Malaysiana 39, 543–547.

Printed in the United States
By Bookmasters